M000159735

"*The Not-So-Intelligent Designer* is a mu
anti-intellectualism is rampant and, sho   ....igii omce
frequently reject evolution. Abby Hafer has that rare ability to communicate
complex scientific ideas in understandable terms for non-scientists, and this
book is sure to enlighten many."

—DAVID NIOSE
Author of *Fighting Back the Right*

"Hafer's book is a valuable contribution to debunking the claims of intelligent
design and the notion of one or more gods intervening in the physics and biology of the real world. She writes in an engaging style that entertains as well as
informs. I enthusiastically recommend it."

—ELLERY SCHEMPP
Plaintiff, *Abington* School District *v. Schempp*;
PhD, Chemical Physics, Brown University

"Intelligent design creationism is a dangerously successful political ideology—
they've passed laws, co-opted high school teachers, and nearly half the population identifies as creationist. *The Not-So-Intelligent Designer* is a guidebook on
how intelligent design fails, from unintelligently designed testes to intelligent
design's unconstitutional religious agenda."

—ZACK KOPPLIN
Organizer of the campaign to repeal the Louisiana Science Education Act,
Baton Rouge, LA

"Hafer's ingenious strategy for dealing with creationists/intelligent design proponents has them by the balls!"

—JOHN W. LOFTUS
Author of *Why I Became An Atheist*

"*The Not-So-Intelligent Designer* is a scholarly book that is accessible and intelli-
gible to the general reader. It is especially a must read for adherents of religious
traditions who embrace modern science. Hafer does a masterful job of defin-
ing science, i.e. a way of knowing characterized by the formulation of a hypothesis, the gathering of evidence, drawing of conclusions, and repeating the
experiment. She shows in a compelling way that the conclusions of advocates
of intelligent design are beyond the purview of science. In fact, she provides a
convincing demonstration of how the battles over intelligent design are over
the nature of science itself."

—LESLIE A. MURAY
Curry College, Milton, MA

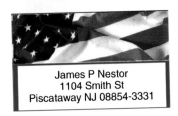
The Not-So-Intelligent Designer

The Not-So-Intelligent Designer

Why Evolution Explains the Human Body and Intelligent Design Does Not

Abby Hafer

Illustrations by
Alexander Winkler

The Lutterworth Press

The Lutterworth Press
P.O. Box 60
Cambridge
CB1 2NT
United Kingdom

www.lutterworth.com
publishing@lutterworth.com

ISBN: 978 0 7188 9420 7

British Library Cataloguing in Publication Data
A record is available from the British Library

First published by The Lutterworth Press, 2016

Copyright © Abby Hafer, 2015

Published by arrangement
with Cascade Books

To Les Muray—scholar, priest, helpful theologian, rock and roller, dedicated humanitarian, world traveler, lover of things Hungarian, serious sports fan, survivor of a life that would have worn out the rest of us long ago, colleague, and friend. Your enthusiastic and knowledgeable support has been a delight.

"Nothing in all the world is more dangerous than sincere ignorance and conscientious stupidity."

—MARTIN LUTHER KING JR.,
STRENGTH TO LOVE

Contents

Acknowledgements

Creating this book has been a long and sometimes meandering process. I had to study the politics and history of the creationism/intelligent design movement as well as the history of denialism in general, learn a whole lot of theology, learn zoology, anatomy, physiology and evolution, do a lot of public speaking, and learn more than I ever knew existed about writing, publishing, photograph hunting and vetting, and biological illustration. Some of these things I started learning long ago, and others have been learned on the fly as I worked on this book. I have been helped in wonderful ways both by friends and by people bestowing kindness upon a stranger. Here is a partial list of those who have helped this book come to fruition. I thank them all for their talent, generosity, and kindness.

THE ENLIGHTENERS: The Reverend Dr. Leslie Muray—who knows nearly everything—helped with theology and also with his enthusiasm for this project. If Les liked the project, it had to count for something. Dr. Ward Holder improved my understanding of Calvinism. Dr. Allan Hunter gave me good advice about writing and publishing. Dr. Keith Wright helped with information theory. Dr. Doug Muder and my sweet husband Alan MacRobert have helped with endless discussions about lots of things. Dr. Roger Hanlon at the Marine Biological Laboratory at Woods Hole helped me understand cuttlefish eyes.

THE HELPERS: Nancy Daugherty bravely did the formatting for this book. Curry College gave me some reduced teaching time so I could work on this book. First Parish in Bedford gave me encouragement, and took an interest in my public speaking. The Secular Students' Alliance enabled me to speak to lots and lots of terrific students all over the country.

Acknowledgements

THE SOURCE FOR GENERAL ANATOMICAL REFERENCE: *Human Anatomy & Physiology*, sixth edition by Elaine N. Marieb (Pearson Benjamin Cummings). Unless noted otherwise, this is the source for my information about human anatomy and physiology. It is well and clearly written, which is a great thing in a textbook.

THE ARTISTS: Oh how I love you. You do wonderful things. Alex Winkler provided all of the drawn illustrations with enthusiasm and skill. He was also very patient, and I thank him for this as well. Gretjen Hargesheimer provided inestimably valuable help in bringing all the photographs to their full potential.

Dr. Gianluca Polgar generously provided me with both photographs and information about mudskippers, which are among my favorite animals. Dr. Alvaro Migotto provided me with a photograph of *Turritopsis nutricula* (the immortal jellyfish)—another of my favorite animals. Dan Norton provided me with a lovely photograph of a pineapple sea cucumber *Thelenota ananas*, and offered me many other pictures as well.

The Natural Arches and Bridges Society (NABS) posts awe-inspiring photographs of (naturally) natural arches and bridges. They also put me in contact with one of their members, Guilain Debossens, who provided me with the beautiful and haunting photograph of the natural arch with tiny feet in the Algerian Sahara.

I also thank the Twin Falls Public Library in Twin Falls, Idaho for being so nice, for allowing me to use their historic photograph of the balanced rock near Castleford, Idaho, and for lovingly preserving the late Clarence Bisbee's wonderful photograph collection.

It has been said that wonder is the basis of worship. It is my hope that through these photographs as well as through my words, people can experience the awe and wonder that is a part of our everyday world, right here, right now.

Chapter 1

Introduction, or, Why Testicles Matter

A few years ago, I realized that the whole intelligent design (ID) controversy is not a scientific issue, but a political one. This goes a long way toward explaining why ID has gotten as far as it has.

ID is not a theory, it is a political pressure group.

Once I realized this, I also realized that my perfect first argument against ID is the male testicle. Why? Because once I mentioned testicles, I knew that people would *pay attention*. Scientific arguments that grab and hold people's attention are what is needed.

The problem is that scientists keep approaching ID as though it were a scientific issue, which it's not. So we make observations, do experiments, and write our papers, showing repeatedly that all the scientific evidence is in favor of evolution. Then we publish our papers in scientific journals where they are read by other scientists. Sometimes we publish wonderful, scholarly books that are also mostly read by other scientists.

I think you see the problem here.

The people who are likely to be persuaded by ID arguments don't read scientific journals, or lengthy books about evolution, *and they never will*. Many of the people who would like to argue in favor of evolution don't read them either. I am not criticizing people who don't read scientific journals. I am criticizing scientists who behave as though talking to other scientists will solve this problem. It won't.

This is why we have libraries full of evidence for evolution and most people don't know it. This means that doing more research won't make a difference.

What will make a difference is understanding that this is a political issue and treating it that way. Political issues require political arguments, and political arguments are different. Political arguments must be short, easy to understand, memorable, and preferably entertaining.

In my case, I also want them to be true.

So when I started looking for new approaches, I knew I had a winner when inspiration hit me in the middle of an Anatomy and Physiology lecture, while I was lecturing about reproductive systems. The male testicle is a great first argument against ID in the human body, and this brings me to the alternative title for this book:

Evolution, Intelligent Design, and Men's Testicles: Why Evolution Explains the Human Body and Intelligent Design Does Not

I understand that this may not be considered an appropriate title for library shelves and book catalogs, but undoubtedly, it would get people's attention. However, once I realized that this was just the sort of thing I needed for a political-style argument, I did what any sensible woman would do under these circumstances. I emailed my minister.

Then my minister did what any *minister* would do under these circumstances, and he told the entire congregation my realization about testicles in that Sunday's sermon.

This is not quite as odd as it sounds. I knew that he, along with many other like-minded ministers, was planning to preach a sermon on Darwin. I thought that my observations about testicles would entertain him and help him understand evolution better. I had forgotten that the Darwin service was that Sunday, and I did not realize that my notes on testicles would lead off the sermon.

You can do this kind of thing if you're a Unitarian.

I later expanded the concept into a booklet, which has been very popular.

There *is* real science here. My point is to show you a number of examples of how the human body is *badly* designed. Given that, is it any wonder then that I started with reproduction?

So without further ado, here are the problems with men's testicles.

Chapter 2

Bad Design—Men's Testicles

The testicles hang outside the body in a sack of skin called the scrotum. Why? Because human body temperature is too hot for sperm production. Having normal body temperature be too hot for sperm production is *bad design.* So the testicles have to hang outside the body in the scrotum, thereby putting a vulnerable organ in a vulnerable place. Putting a valuable and vulnerable organ in such a vulnerable location is *bad design.* Men are put to all sorts of inconvenience and risk severe pain and worse because of this unfortunate positioning. One would think that God could do better.

Here's a picture of how this is all put together.

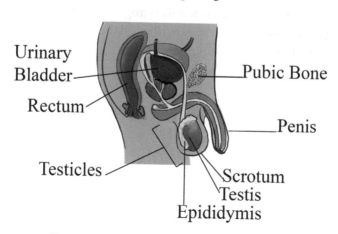

Figure 2.1   Reproductive system of the male human.

Notice how the testicles, with their inability to be warm and productive at the same time, hang outside, while all the abdominal organs are safely tucked up inside, out of harm's way. Our cold-blooded relatives don't have this problem, and their sperm-making equipment is safely inside them. If you don't believe me, try to find the balls on a frog.

You won't manage it unless you do a dissection.

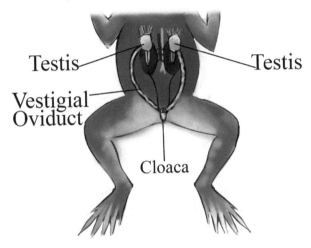

Figure 2.2   Reproductive system of the male frog.

Here's a picture of a frog's insides. See? The frog's testicles are safe inside him, where a vulnerable organ ought to be.

Does this mean that the Creator likes frogs better than men, or does it mean that as humans evolved, the ones who had their balls hanging outside reproduced better? You decide.

Chapter 3

What Is Intelligent Design, and What Does It Have to Do With Men's Testicles?

So, what do male testicles have to do with ID? Little did we realize that this would become one of the central questions of modern science.

Proponents of ID insist that biological organisms everywhere, including human beings, show unmistakable signs of having been designed by an intelligent Creator, rather than having evolved through natural selection.

But if testicles were designed, then one wonders why God didn't protect them better. Couldn't the Designer have put them inside the body, or encased them in bone, or at least put some bubble wrap around them? Is this the best that the Designer can do?

ID is a very important idea. Its advocates have support from numerous presidential candidates, some members of Congress, a few United States governors, and many state legislatures. They are the people responsible for the famous court case called *Kitzmiller v. Dover Area School District.*

They think that educational policy and textbooks should be changed to reflect their views. Who are these people, what do they believe, and how did they get to be so powerful?

WHAT IS INTELLIGENT DESIGN?

ID is the idea that biological organisms have come about due to the deliberate work of an intelligent Creator. ID says that the Creator's "signature" can be seen in the way we are put together.

It further argues that new species cannot come about through evolution by natural selection, and must be the work of a Designer. This means that human beings, which are a separate species, as well as all other creatures, are considered to be the products of intelligent design.

This idea admits that evolution by natural selection can modify existing features, but only within species that already exist, and that new features are also the work of a Designer.

ID's proponents insist that it is as valid a scientific theory as evolution by natural selection and that it therefore must be taught alongside evolution in science classes in public schools. ID's proponents are also unique in the history of science for insisting that their views be written into science textbooks before a single experiment has been done.

Why Does This Matter?

"But wait a minute!" I hear you say. "Wasn't this settled by the Scopes trial in 1925? And by the Dover, Pennsylvania school trial of 2005?" Unfortunately, the problem is that the ID folks are a political lobby, just like the tobacco lobby. They don't give up, because they want you to buy their product, no matter what.

What's ID's product? Religious indoctrination. As the judge in the Dover school trial said,

> [W]e conclude that the religious nature of ID [intelligent design] would be readily apparent to an objective observer, adult or child.[1]

ID proponents want your kids and my kids to grow up being taught the ID proponents' version of religion in public school. At your expense, since you pay taxes. Here is a quote from their strategy document, called the Wedge Strategy:

1. *Kitzmiller v. Dover Area School District,* Judge John E. Jones III's decision, 24.

> Design theory promises to reverse the stifling dominance of the materialist worldview, and to replace it with a science consonant with Christian and theistic convictions.[2]

By "materialist," here, they don't mean buying stuff. They mean believing in facts and evidence about the material world. In other words, scientific facts. They want to squash science as a method of investigation, which obtains facts about the material world by investigating it using material means.

What's more, when ID promoters talk about wishing to replace modern science with "a science consonant with Christian and theistic convictions," you need to know that "theistic convictions" means that God created and rules the world, and no explanation is acceptable if it doesn't put God first. So ID proponents don't like science, because it doesn't invoke God for its explanations. And when ID proponents talk about theism and God, they specifically mean a conservative Christian version of God. They may say otherwise when they talk to the press, but their writings reveal their insistence on the conservative Christian world view, which they think should take over science.

ID proponents want everyone in the US, by way of public schools, to be taught that the actual facts about the material world don't exist, or shouldn't. Instead, they simply want to tell you what you have to believe, regardless of any factual basis. In other words, if they invent it, you have to believe it.

Attacking the teaching of evolution is simply their way of getting into the American school system. They try to convince politicians that what they are saying is science, not religion, so that then they can force their way into American public education, and then expand from there. They see this as a political fight, and are using political means to fight it.

Who Is Promoting Intelligent Design?

Although it presents itself as a grassroots concern, ID promotion is actually a well-run and well-funded political operation. One of the places that pushes it very hard is the Discovery Institute. The Discovery Institute is located in Seattle, Washington. It has received a great deal of funding from multimillionaire Howard F. Ahmanson Jr. Mr. Ahmanson was quoted in the *Orange County Register* in 1985 as saying, "My goal is the total integration

---

2. *The Wedge Strategy: Five Year Strategic Plan Summary,* 4. This document may be accessed at http://ncse.com/creationism/general/wedge-document.

of biblical law into our lives."[3] Other wealthy conservative and religious entities also contribute to the Discovery Institute.

The Discovery Institute and its subsidiary, the Center for Science and Culture, has a long list of fellows, directors, program advisors, and program directors. Many of these people make handsome salaries for promoting ID. They want to replace scientific investigation with the words "God did it." They think that this is an adequate and even preferable explanation for everything, despite the fact that I have never seen a successful satellite launch that based its knowledge of physics on biblical writings.

Here is another quote from the Wedge Strategy, describing the goal of the Discovery Institute:

> Discovery Institute's Center for the Renewal of Science and Culture seeks nothing less than the overthrow of materialism and its cultural legacies.[4]

Remember that when they say "materialism," what they mean is science. So ID is very well funded, well organized, very determined, and they want to indoctrinate American children and American society with their antiscientific rubbish, at taxpayer expense.

What Is the Evidence for Intelligent Design?

Proponents of ID have a wide range of viewpoints. Some point to the Cambrian explosion, a time when many species came into existence, as evidence for ID. They say that all these species couldn't have come into existence without help from somebody.

ID promoters have a number of concepts that they work with. These include *irreducible complexity*, *specified complexity*, and the *design inference*. Here's what those terms mean.

IRREDUCIBLE COMPLEXITY means that some people believe that certain biological structures or systems are too complex to have evolved from similar structures or systems in simpler organisms.

So a feature or system is irreducibly complex if it has many distinct parts, all of which are necessary for its proper functioning. If any single part

3. Cited in Kerwin Lee Klein, *From History to Theory* (Berkeley, CA: University of California Press, 2011) 153.

4. See http://ncse.com/creationism/general/wedge-document.

is removed from this system, then it no longer functions properly, and this makes it irreducibly complex.

Here is what Michael Behe, the primary proponent of irreducible complexity, has to say on the subject: "Irreducible complexity is just a fancy phrase I use to mean a single system which is composed of several interacting parts, and where the removal of any one of the parts causes the system to cease functioning."[5] Behe is confusing destruction and simplicity. He doesn't say "irreducible complexity is just a fancy phrase I use to mean something that couldn't have evolved from something simpler." Rather, he proposes looking at a living, functioning system and removing parts from it. If the system then ceases to function, then Behe wishes you to believe that it cannot have evolved into existence. So, for instance, if you chop a dog's head off and it dies, that proves it couldn't have evolved from a simpler organism.

The usual examples given for irreducible complexity are the human blood clotting sequence, the bacterial flagellum, and the human eye. I will talk about these in Chapters 18, 20, and 21.

SPECIFIED COMPLEXITY insists that specific complex patterns in their current forms, such as some biological systems, are unlikely to have evolved through random mutation. Some ID promoters insist that specified complexity proves that evolution could not have produced new species.

DESIGN INFERENCE is a similar argument. It also insists that the probability that specific complex patterns evolved through random mutations is so low as to be impossible. It says that specific complex patterns are seen in many current-day biological systems. This then leads ID promoters to insist that a Designer must have brought these improbable outcomes into being.

Unfortunately, this is kind of like shooting an arrow randomly, and then when it lands somewhere, painting a bull's-eye around where it hit, and then saying "God made me hit a bull's-eye! The odds are too infinitely small for me to have hit that bull's-eye by chance! Therefore, God directed my arrow."

5. Behe, "Evidence for Intelligent Design from Biochemistry," http://www.discovery.org/a/51.

What Do They Really Believe?

Some ID proponents insist that the earth is at most 10,000 years old. Others agree with modern geology and say that the earth is approximately 4.56 billion years old. Most ID promoters insist that ID is not religion. Others insist that the Designer is God, and by God, they mean the Christian God.

In general, the range of viewpoints among ID proponents is very wide. There's no agreed-upon theory of who the Designer is, when and how the Design was implemented, which interspecies barriers are inviolable, how new species are created, or how new features are put into existing organisms.

In fact the only thing ID proponents have in common besides, in many cases, fat paychecks from the Discovery Institute, is that they insist that their version of reality must be taught in public schools at taxpayer expense.

Public Education

The basic argument that ID's supporters use is that they have expressed doubt about evolution, and this therefore means that their viewpoint must be taught as science in science classrooms.

One strong piece of evidence that ID is not science is that its promoters insist on its being written into textbooks and taught in public schools before they have conducted a single experiment.

ID promoters ignore the fact that having a few people with a contrary viewpoint does not amount to a serious controversy. Scientists around the world accept the overwhelming evidence that biological organisms evolved through natural selection. A few crackpots claiming something else does not amount to an important controversy.

Of course, what they are really trying to do is teach their particular religion in American public schools at taxpayer expense. They pretend that it's science, but by their own admission, their stated goal is to destroy science. They wish to insert their religion into public schools, so that all children are indoctrinated with their religion. All paid for by American taxpayers.

Politics

ID promoters are very energetic in their pursuit of public education. They intend to win their fight not by proving their claims scientifically, but by winning in the court of public opinion. To further their goals, they get candidates to run for school boards, without telling the public what their beliefs are. Since most schools' boards need people to volunteer to be members, these stealthy candidates can get themselves elected. When they achieve a majority, they then announce that ID will be taught in public schools. This is what happened in Dover, Pennsylvania.

ID promoters also lobby politicians and political institutions at all other levels of government. So, for instance, presidential candidate and then-senator Rick Santorum amended an educational funding act to encourage presenting evolution as "continuing controversy."[6] This has been used by ID proponents throughout the United States as an excuse for proposing legislation that encourages teaching ID as science in public school science classrooms. Since then Louisiana has passed the ID-friendly "Louisiana Science Education Act," which claims only to wish to promote critical thinking, but manages only to include in its list of subjects worthy of critical thought the fields of "evolution, the origins of life, global warming, and human cloning."[7] Louisiana Governor Bobby Jindal signed this into law in 2008.

As of 2012, Tennessee also enacted a law that allows the teaching of intelligent design/creationism in public schools.[8]

In 2014, the states of Ohio,[9] South Dakota,[10] Missouri,[11] Virginia,[12] and Oklahoma[13] all had bills in their state legislatures that would have

6. The National Center for Science Education, "Is There a Federal Mandate to Teach Intelligent Design Creationism?", http://ncse.com/taking-action/analysis-santorum-language.

7. See the Louisiana Science Education Act, http://www.legis.state.la.us/lss/lss.asp?doc=631000.

8. House Bill 368, Senate Bill 893, http://www.capitol.tn.gov/Bills/107/Bill/HB0368.pdf.

9. House Bill 597, http://www.legislature.state.oh.us/bills.cfm?ID=130_HB_597.

10. Senate Bill 112, http://legis.sd.gov/Legislative_Session/Bills/Bill.aspx?file=SB112P.htm&&Session=2004&cookieCheck=true.

11. House Bill 1472, http://www.house.mo.gov/billsummary.aspx?bill=HB1472&year=2014&code=R.

12. House Bill 207, http://legl.state.va.us/cgi-bin/legp504.exe?141+ful+HB207.

13. House Bill 1674, http://www.oklegislature.gov/BillInfo.aspx?Bill=HB1674&

allowed the teaching of intelligent design/creationism as science, in American public schools.

This is what I mean by winning in the court of public opinion.

They don't do scientific experiments to show the truth of their claims, they just lobby politicians whether their claims are true or not. In fact, they repeat claims that have been publicly exposed as being untrue. They also make up new claims very easily, since they don't feel the need to base their claims on facts or experimentation.

Unfortunately, this means that no matter how many of their silly claims serious scientists defeat, ID promoters simply repeat their untrue claims, and crank out more of them.

This is a politically expedient approach for ID to take because it's easier to come up with specious claims than it is to do real scientific experiments. Nonetheless, even knowing that they may come up with five new unsubstantiated new claims next week, I will address some of their current claims in the following chapters.

But for now, what about the big one? Their major claim is that we are intelligently designed. Let's get back to having fun. Let's talk about testicles. Again.

Session=1300. Senate Bill 1765, http://www.oklegislature.gov/BillInfo.aspx?Billsb1765& Session=1400.

Chapter 4

Testicles, Part II

"But wait!" I hear you cry. "What if warm body temperature and sperm production just *can't* go together? What if warm-blooded animals really have to have their balls on the outside?" Well, here's one solution.

Convertibles—Not Just for Cars

You'd think that the Designer would try to give us humans, who were made in his image, the best deal that he could manage, wouldn't you? But he didn't.

Some warm-blooded animals have testicles that hang outside the body when this is needed, but which can be pulled up inside the body out of harm's way when this is the safer option.

Actually, lots of mammals get this deal. For instance, rats, mice, rabbits, hares, cavies, and guinea pigs can all do this. So when they want to breed, the testicles can hang outside and stay cool. But when they're not breeding, they can be pulled up inside for safekeeping.

Think of the convenience! Many men would kill for this option. First off, men could control their fertility easily, in a convenient and nonpermanent way. They could pull their testicles in most of the time, which would both keep them safe, and prevent unintended pregnancies. No guesswork, no rhythm method, no trips to the drugstore, no surgeries.

But when they wanted to breed, they could pull their testicles out for a few days, make some fresh sperm (human sperm are made new every day, by the way), and then be ready to breed, when they want, and not when they don't want. Rabbits got this deal. Rats got this deal. But not us.

So who does the Designer like better—mice, or men?

Warm Body Temperature and Sperm Production—It Can Be Done

To make matters worse, in turns out that warm body temperature and sperm production really can go together—just not in humans. For instance, elephants, our fellow warm-blooded mammals, have their testicles inside them.

Here's an illustration of an elephant's insides:

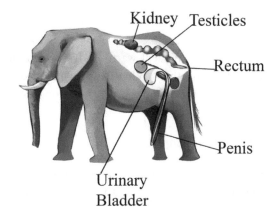

Figure 4.1   Elephants have their testicles deep within their bodies.

Armadillos, sloths, hyraxes, echidnas, platypuses, seals, whales, dolphins, manatees, and rhinoceroses also have internal testes.

So the Creator gave elephants and rhinoceroses protection for their testicles, but didn't give it to the much smaller and more vulnerable human beings.

Birds Do It

What's more, the argument that warm body temperature and sperm production just can't be made to work together would also work better if birds didn't exist, but they do. Birds are warm-blooded too. In fact, they have higher body temperatures than we do. But, like frogs, and elephants, their testicles are safely tucked up inside them. If you don't believe me, try to find the balls on your Thanksgiving turkey.

Again, you won't manage it unless you do a dissection—before cooking.

Here is a picture of a bird's insides:

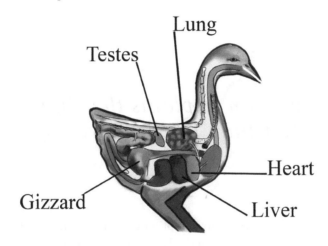

Figure 4.2   Birds do too.

Once again, the bird's testicles are safe inside him, where a vulnerable organ ought to be. So—does the Designer like birds better than us? Or did God make birds in his image? In other words, is the Designer a turkey?

Chapter 5

Why Evolution Explains the Human Body Better than Intelligent Design Does

Here is my primary reason why evolution explains the human body so much better than ID does: the standards for evolution are much *lower*.

The standard for systems that evolve is "good enough to not cause death before reproduction, too much of the time."

The standard for ID is "designed by an infallible Creator." You can see the difference here.

Believe me, evolution explains human anatomy far better than any notion of a good designer. Human bodies are just too badly put together to stand up to even reasonable design specifications, much less infallible ones.

Chapter 6

Intelligent Design According to Its Believers: Is Intelligent Design the Same as Creationism, and Is Intelligent Design Religion?

People get confused about ID because it is a very broad subject. In this chapter I will discuss what it is according to its believers, how it relates to creationism, and how it relates to religion.

Intelligent Design According to Its Believers

ID asserts that sometime in the past, in ways not described, some singular being (possibly God) created life, and created all the species of biological organisms on the planet. However, there is a wide diversity of opinions among the proponents of ID as to exactly what ID really is.

Here are the opinions about ID from many of ID's most vocal promoters, in their own words.

PHILLIP E. JOHNSON is the founding father of the modern ID movement. He believes that ID is correct because he does not like the idea that science restricts itself to looking for natural causes for natural phenomena.

> We are taking an intuition most people have (the belief in God) and making it a scientific and academic enterprise. We are

removing the most important cultural roadblock to accepting the role of God as creator.[1]

However, Johnson, who is a law professor rather than a scientist, also publicly insists that AIDS is not caused by the HIV virus. Johnson is retired from the University of California at Berkeley Law School, and is the co-founder and program advisor for the Center for Science and Culture at the Discovery Institute.

DR. WILLIAM DEMBSKI, a prominent ID promoter, insists that ID is not religion. He makes this a contrast between ID and creationism, which he says *is* religion:

> Intelligent design, by contrast [to creationism], places no such requirement on any designing intelligence responsible for cosmological fine-tuning or biological complexity. It simply argues that certain finite material objects exhibit patterns that convincingly point to an intelligent cause. But the nature of that cause—whether it is one or many, whether it is a part of or separate from the world, and even whether it is good or evil—simply do not fall within intelligent design's purview.[2]

Though sometimes he says that ID is religion after all. Specifically, it is Christianity.

> It is no accident that the first thing the Bible teaches is creation. Creation implies purpose. Because we are created, there is a purpose for our existence, for the family, for work, for sex, and for how we ought to live. Creation by a loving God is our *origin*.[3]

Dr. Dembski is a philosopher and mathematician. He is a Senior Fellow at the Discovery Institute's Center for Science and Culture and has recently been a research professor in philosophy at Southwestern Baptist Theological Seminary.

1. Quoted in Teresa Watanabe, "Enlisting Science to Find the Fingerprints of a Creator," *Los Angeles Times*, March 25, 2001.

2. Dembski, *Expert Witness Report: The Scientific Status of Intelligent Design*, www.designinference.com/.../2005.09.Expert_Report_Dembski.pdf.

3. William Dembski and Sean McDowell, *Understanding Intelligent Design: Everything You Need to Know in Plain Language* (Eugene, OR: Harvest House, 2008), 18.

Intelligent Design According to Its Believers

DR. JONATHAN WELLS (JOHN CORRIGAN WELLS) insists that Intelligent Design is science.

> ID maintains that it is possible to infer from empirical evidence that some features of the natural world are best explained by an intelligent cause rather than unguided natural processes.
>
> Three things are noteworthy about this description of ID. First, design is inferred from evidence, not deduced from scripture or religious doctrines. All of us make design inferences every day, often unconsciously. ID attempts to formulate our everyday logic in terms rigorous enough to warrant inferences from the evidence in nature. This is clearly not the same as biblical creationism.[4]

Though he arrived at this conclusion though prayer, rather than because there was evidence that convinced him, and because of advice from Reverend Sun Myung Moon.

> At the end of the Washington Monument rally in September, 1976, I was admitted to the second entering class at [Moon's] Unification Theological Seminary. During the next two years, I took a long prayer walk every evening. I asked God what He wanted me to do with my life, and the answer came not only through my prayers, but also through Father's [Moon's] many talks to us, and through my studies. Father encouraged us to set our sights high and accomplish great things.
>
> He also spoke out against the evils in the world; among them, he frequently criticized Darwin's theory that living things originated without God's purposeful, creative activity. My studies included modern theologians who took Darwinism for granted and thus saw no room for God's involvement in nature or history; in the process, they re-interpreted the fall, the incarnation, and even God as products of human imagination.
>
> Father's words, my studies, and my prayers convinced me that I should devote my life to destroying Darwinism, just as many of my fellow Unificationists had already devoted their lives to destroying Marxism. When Father chose me (along with about a dozen other seminary graduates) to enter a Ph.D. program in 1978, I welcomed the opportunity to prepare myself for battle.[5]

4. Wells, "Give Me That Old Time Evolution: A Response to the *New Republic,*" http://www.discovery.org/a/2933.

5. Wells, "Darwinism: Why I Went for a Second Ph.D.," http://www.tparents.org/

He admits that the earth is many billions of years old.

> Since 1859, however, many Precambrian fossils have been found, including microfossils of single-celled bacteria in rocks more than three billion years old. In addition, multicellular Precambrian fossils have been found in the Ediacara Hills of Australia, though there is continuing debate over whether any—or how many—of the Ediacaran fossils were animals, or what relationship—if any— they had to the Cambrian phyla. In 1998, Cambridge University paleobiologist Simon Conway Morris (who is featured in the film "Darwin's Dilemma") wrote, 'Apart from the few Ediacaran survivors . . . there seems to be a sharp demarcation between the strange world of Ediacaran life and the relatively familiar Cambrian fossils' (*Crucible of Creation*, 30).[6]

Dr. Wells has PhDs in both molecular and cell biology and in religious studies. Dr. Wells also denies that HIV is the cause of AIDS. He is a senior fellow at the Discovery Institute's Center for Science and Culture.

Dr. Paul A. Nelson is a Young Earth creationist. Young Earth creationists generally believe that the earth is no more than 10,000 years old, at most. However, Dr. Nelson isn't quite sure what he means by "young."

> We hold the view of *recent* or so-called *young earth* creation. Unfortunately, neither "young earth" nor "recent" is satisfactory as a descriptive adjective.[7]

Though he acknowledges that the scientific evidence is against this view.

> Natural science at the moment seems to overwhelmingly point to an old cosmos. Though creationist scientists have suggested some evidence for a recent cosmos, none are widely accepted as true. It is safe to say that most recent creationists are motivated by religious concerns.[8]

Dr. Nelson also believes that there is not yet a genuine theory of intelligent design.

---

library/unification/talks/wells/darwin.htm.

6. Wells, *Deepening Darwin's Dilemma,* Discovery Institute September 16, 2009.

7. Paul Nelson and John Mark Reynolds, "Young Earth Creationism," in J. P. Moreland and John Mark Reynolds, eds., *Three Views on Creation and Evolution* (Grand Rapids: Zondervan, 1999), 41.

8. Ibid., 49.

> Easily the biggest challenge facing the ID community is to develop a full-fledged theory of biological design. We don't have such a theory right now, and that's a problem. Without a theory, it's very hard to know where to direct your research focus. Right now, we've got a bag of powerful intuitions, and a handful of notions such as 'irreducible complexity' and 'specified complexity'—but, as yet, no general theory of biological design.[9]

Dr. Nelson is a philosopher and a fellow at the Discovery Institute's Center for Science and Culture.

DR. MICHAEL BEHE accepts the idea of common descent among organisms.

> The word "evolution" carries many associations. Usually it means common descent—the idea that all organisms living and dead are related by common ancestry. I have no quarrel with the idea of common descent, and continue to think it explains similarities among species. By itself, however, common descent doesn't explain the vast differences among species.[10]

He also accepts that the universe is billions of years old.

> For the record, I have no reason to doubt that the universe is the billions of years old that physicists say it is. Further, I find the idea of common descent (that all organisms share a common ancestor) fairly convincing, and have no particular reason to doubt it.[11]

And he accepts that humans and chimpanzees share a common ancestor.

> For example, both humans and chimps have a broken copy of a gene that in other mammals helps make vitamin C. . . . It's hard to imagine how there could be stronger evidence for common ancestry of chimps and humans. . . . Despite some remaining puzzles, there's no reason to doubt that Darwin had this point right, that all creatures on earth are biological relatives.[12]

9. Nelson, "The Measure of Design," *Touchstone,* July 8, 2004, 64–65.

10. Behe, "Darwin Under the Microscope," *New York Times,* October 29, 1996.

11. Behe, *Darwin's Black Box: The Biochemical Challenge to Evolution* (New York: Free Press, 2006), 5–6.

12. Behe, *The Edge of Evolution: The Search for the Limits of Darwinism* (New York: Free Press, 2008), 71–72.

Dr. Behe is a biochemist who is a professor of biochemistry at Lehigh University and is a senior fellow at Discovery Institute's Center for Science and Culture.

A WIDE DIVERSITY OF OPINIONS . . .

So actually, when I said that there is a wide diversity of opinions among the proponents of ID as to exactly what it is, what I really meant is that these guys couldn't agree that gravity makes things fall down.

Although they may have other explanations, their willingness to simultaneously assert and deny that ID is religion at least appears fundamentally dishonest.

. . . BUT ONE AGENDA

The only thing they do agree on, however, is that they want to teach ID to your children as science, and they expect you to pay for it. They are well funded, well organized, and have no idea what they're talking about. However, even though they can't agree on what ID is, they insist that it be in science textbooks, and be taught in public schools.

They also know how they want to accomplish this. They planned it all out in 1998, when they produced an action plan for forcing ID into public schools. They called this action plan the Wedge Strategy. I discuss the Wedge Strategy in Chapter 7.

Is Intelligent Design the Same as Creationism?

ID proponents claim that ID is different from creationism. Here's what intelligent design has to say about itself and creationism.

> Creationism typically starts with a religious text and tries to see how the findings of science can be reconciled to it. Intelligent design starts with the empirical evidence of nature and seeks to ascertain what inferences can be drawn from that evidence. Unlike creationism, the scientific theory of intelligent design does not claim that modern biology can identify whether the intelligent cause detected through science is supernatural.[13]

13. From intelligentdesign.org, February 24, 2010. Intelligentdesign.org is a part of the Center for Science and Culture, which is a part of the Discovery Institute.

So according to ID proponents themselves, creationism is based on religion. ID, on the other hand, they say, looks at the real world. But they don't do experiments. Note the lack of the words "experimental evidence" anywhere in their definition of ID.

What They Say and What They Really Mean

Creationism actually claims that it isn't religion, but it is. ID also claims it isn't religion, but it is. However, they do not state exactly the same thing all the time. So in that way creationism and ID are different. However, they were invented by the same people, for the same purpose.

In fact, they are so close that the same textbook was written for both of them.

Here's what I mean.

"Creationists," "Design Proponents," and "Cdesign Proponentsists"

In 1987, The Supreme Court decided that teaching creationism in public schools violated the Establishment Clause of the United States constitution. The case was called *Edwards v. Aguillard.*

The Establishment Clause is the one saying that we won't have any official state religion. This means that we therefore shouldn't advocate a particular religion in public schools. The Supreme Court found that creationism qualifies as religion, and therefore can't be taught in American public schools.

The creationists had written a textbook espousing creationism called *Of Pandas and People.* They wanted public schools to adopt it. Unfortunately for the creationists, once creationism was found to be religion by the Supreme Court, public schools weren't going to buy it.

So they went back to the book, and took out all the words that sounded like creationism, and put in words that sounded like intelligent design. For instance, the word *creationism* was changed to *intelligent design*, and the word *creationists* was changed to *design proponents*. Except that in one case they were sloppy. The word *creationists* was not completely removed before the words *design proponents* were put in, and the book said *cdesign proponentsists*. This proved that ID and creationism are essentially the same thing, being pushed by the same people, for the same reasons.

Here are copies of the two texts from the website for the National Center for Science Education:

```
                              3-40
        The basic metabolic pathways (reaction chains) of nearly all
   organisms are the same.  Is this because of descent from a common
   ancestor, or because only these pathways (and their variations)
   can sustain life?  Evolutionists think the former is correct,
   creationists accept the latter view.  Creationists reason as
```

Figure 6.1   *Of Pandas and People* (1987, "creationist" version), ch. 3, p. 40.

```
        The basic metabolic pathways (reaction chains) of nearly all
   organisms are the same.  Is this because of descent from a common
   ancestor, or because only these pathways (and their variations)
   can sustain life?  Evolutionists think the former is correct,
   cdesign proponentsists accept the latter view.  Design proponents
```

Figure 6.2   *Of Pandas and People* (1987, "intelligent design" version), ch. 3, p. 41.

You can see that the two texts are exactly the same, except that the word *creationists* was replaced with the words *design proponents*—and that sometimes they didn't do a very good job of swapping the words.

"Cdesign Proponentsists"—the Transitional Species!

Opponents of evolution often point to the fossil record and say, incorrectly, that there isn't any evidence in the fossil record of organisms slowly transitioning over time from one species to another. The in-between organisms, that have some of the characteristics of the older organisms, and some characteristics of the later organism that the species eventually becomes, are known as transitional species. The fossil record is full of them, but design proponents ignore this.

However, the written record is very clear. "Cdesign proponentsists" is the transitional species between creationism and ID, and shows that creationism is the institutional ancestor of ID.

Kitzmiller v. Dover Area School District

The book *Of Pandas and People* was used as part of the evidence in a later court case when design proponents again wanted to insist on teaching ID in public schools. This time the schools were in Dover, Pennsylvania and the year was 2005.

This case was called *Kitzmiller v. Dover Area School District*. In it, Judge John E. Jones III found that ID and creationism are fundamentally the same thing, and that therefore since creationism is religion, then ID is too. It therefore can't be taught in public schools. Here's what the judge said:

> As Plaintiffs meticulously and effectively presented to the Court, *Pandas* went through many drafts, several of which were completed prior to and some after the Supreme Court's decision in Edwards, which held that the Constitution forbids teaching creationism as science. By comparing the pre and post Edwards drafts of *Pandas*, three astonishing points emerge: (1) the definition for creation science in early drafts is identical to the definition of ID; (2) cognates of the word creation (creationism and creationist), which appeared approximately 150 times, were deliberately and systematically replaced with the phrase ID; and (3) the changes occurred shortly after the Supreme Court held that creation science is religious and cannot be taught in public school science classes in Edwards. This word substitution is telling, significant, and reveals that a purposeful change of words was effected without any corresponding change in content . . . . The weight of the evidence clearly demonstrates, as noted, that the systemic change from "creation" to "intelligent design" occurred sometime in 1987, after the Supreme Court's important Edwards decision.[14]

These days, the ID folks try hard to distance themselves from creationism in public, but not in private. Remember that Dr. Paul Nelson, who pushes ID, is a Young Earth creationist.

Dr. Nelson's Young Earth creationists believe that the earth is at most 10,000 years old, and that all biological organisms, as well as the earth and the universe, were created as direct acts of the Abrahamic God.

14. Judge John E. Jones III, *Kitzmiller v. Dover Area School District*.

Dr. Nelson continues to be a fellow with the Discovery Institute's Center for Science and Culture, so obviously, the ID lobby likes him pretty well.

What's more, the book *Of Pandas and People* now has a new edition, retitled as *Design of Life*, and the new authors listed on the cover are William A. Dembski and Jonathan Wells, both of whom are prominent ID proponents.

Of course, Jonathan Wells also denies that the HIV virus causes AIDS, so it's not clear how reliable he is as a science textbook author.

IS INTELLIGENT DESIGN RELIGION?

It is clear that ID starts from an overtly conservative Christian viewpoint, and seeks to promote this as science. Although ID promoters shy away from talking about their God in their official definition of themselves, they talk long and loud about God and intelligent design elsewhere.

To quote Phillip E. Johnson, the father of ID:

> Our strategy has been to change the subject a bit so that we can get the issue of intelligent design, which really means the reality of God, before the academic world and into the schools.
>
> —American Family Radio, January 10, 2003 broadcast

And also by Johnson:

> This isn't really, and never has been, a debate about science. It's about religion and philosophy.[15]

What's more, Dr. William Dembski, another famous ID promoter, has had the following things to say in a book promoting ID:

> God the Father creates through the Son in the power of the Holy Spirit. By a free act that mirrors the intratriune personal relations, God creates a world of finite creatures (which include physical as well as spiritual beings, humans as well as angels).[16]

---

15. Johnson, "Witnesses for the Prosecution," *World Magazine*, November 30, 1996, 18.

16. Dembski, *The Design Revolution* (Downers Grove, IL: InterVarsity, 2009), 175.

And:

> "Christian theology, properly so-called, regards the doctrine of *creation ex nihilo* or *creation from no preexisting stuff* as nonnegotiable."[17]

And this is in a book about ID, a movement that claims publicly to be nonreligious!

Other books by Dr. Dembski spell it right out in the book description itself, such as this one. Here's the title: *Understanding Intelligent Design: Everything You Need to Know in Plain Language*. Here's the book's description as written by Dr. Dembski himself on his own website: *"A user-friendly introduction to ID for Christian young people."*[18]

This makes the connection between unconcealed Christian religion and ID perfectly plain.

They Publish in a Religious Journal and through a Religious Publisher

To make things still clearer, the people who write about ID often publish their essays in a journal called *Touchstone*. The full title for *Touchstone* is *Touchstone: A Journal of Mere Christianity*. This tells us that not only are these folks pushing Christian religion; they've got a chip on their shoulders about it as well (and they like C. S. Lewis).

They publish many of their books through a Christian religious press called InterVarsity Press. According to its own website, "InterVarsity Press is a Christian publishing company dedicated to serving the university, the church and the world."

Make no mistake. The ID lobby is a Christian religious lobby.

So there you have it. ID promoters really do assert that their "intelligent designer" is God, and they really do know that ID is not science, and they do acknowledge that their aim is to insert it into American public schools, whether the Constitution likes it or not.

And finally, their choice of words gives away their intention from the outset. They assume that they will find the Christian God. After all, ID promoters always refer to a singular "Designer." Why do they assume that there's only one Designer? Why not many? Why not a "Design Team"? Why not competing "Designers"? But no. They always refer to *"the* Designer"—not

---

17. Ibid., 174.

18. See http://www.designinference.com/.

"the Designers," or "the Committee." In other words, they are assuming, based on nothing, that their Designer is one, monotheistic entity. In short, the Judeo-Christian God.

Now in case you had any doubts as to what ID promoters really want, or whether or not it really has a religious agenda, read about when they tipped their hand in the secret, infamous document titled *The Wedge Strategy*.

Chapter 7

The Infamous Wedge Strategy

The Wedge Strategy is the title of a document, and it is probably the first document in history to propose defeating science by using a public relations campaign. Written in 1998, *The Wedge Strategy* is the document that the Discovery Institute wishes you didn't know about. In fact, it was marked "Top Secret" and "Not for Distribution." The only reason we know about it is because it was leaked to the web in 1999.[1]

Basically, the Discovery Institute decided that it needed to take over American science, education, and society. It wants to replace them with something that its members approve of and can be in charge of.

So in 1998, they created *The Wedge Strategy*. This document plans out how they are going to do this.

They intend to force this takeover using ID. In fact, they refer to their strategy as the "Wedge Strategy" because they want to use ID as the thin end of a wedge, like an axe, that would split open science, and create an opening for their religious ideas as a replacement for science. If you think that I am stating the case too strongly, here is the cover they used for their Wedge Strategy document. It is a scene of biblical creation, with a superimposed triangle, representing their axe-like "wedge."

---

1. The entire *Wedge* document can be found online at http://ncse.com/files/pub/creationism/The_Wedge_Strategy.pdf.

Figure 7.1   Copy of the cover of *The Wedge Strategy*.

In *The Wedge Strategy*, the Discovery Institute stated its primary aim, which is to "defeat scientific materialism." Remember, by materialism, they don't mean buying lots of stuff. They mean believing in the material world. If you only accept evidence from the material world, evidence like things you can actually see and hear, then according to ID proponents you are a "materialist."

The Discovery Institute doesn't like "materialism." However, science requires that scientists only accept evidence from the material world when they are doing their research. This specific requirement is what has allowed science to advance. In the last 400 years, our understanding of the natural world has progressed to a previously unimaginable degree as a direct result of this requirement. In fact, science is the all-time undisputed champion at finding out material facts about the material world as direct result of this requirement.

This requirement has made scientists look for real-world explanations for natural phenomena, which we have then found. This change in our knowledge is so large and magnificent that the method that we used to achieve it should be respected, not dismantled. You can't get the same results with different means.

Falling back on a deity may seem like a reasonable idea at times, but it doesn't produce useful results.

Rejecting supernatural explanations and seeking real ones is what has allowed us to find those real explanations. Every time.

For example, think about AIDS. After AIDS was discovered, many people were happy to call it the wrath of a deity. Fortunately, scientists did the research and discovered that the real cause of AIDS is HIV. Because of that discovery, we are now able to successfully treat people with AIDS. That's how science progresses—by seeking reality-based answers rather than fake but easy ones offered by supernatural explanations.

Supernatural explanations would stop science in its tracks, and this is what the Discovery Institute wants.

So let's not pretend that the Discovery Institute is doing science. It is doing the opposite. It wants to replace science. It doesn't like "materialism," so it doesn't like science, because science only uses evidence from the material world.

The Discovery Institute's stated goal is "to replace materialistic explanations with the theistic understanding that nature and human beings are created by God." It is not content with having God discussed in churches and in private contemplation. It wants to shove its idea of God into science, where it would damage people's health and happiness.

So forget any notion that ID is science. The entire aim of its existence is to obliterate science.

In *The Wedge Strategy*, the Discovery Institute outlines how it's going to do it. It has five-year plans, and twenty-year plans.

It talks about ID taking over, but not about what ID actually is. It talks about research, but not experiments. It talks about having numerous research articles in scientific journals, but that didn't happen. It talks about "persuading" people, but not about accumulating evidence.

It talks about how bad its authors think science is, and why they think that society is bad because of science's influence.

It also talks about how science made people think that life in the world could be better, which the ID promoters branded as "utopianism." They state that applying scientific knowledge to human problems is wrong, so it appears that intelligent design promoters are against things like the Clean Air Act and the Clean Water Act, because these use scientific knowledge to create a better life for people here on Earth.

And they talk at length, and more length, and more length, about how they will run their public relations campaign. *Not* how they will accumulate evidence for ID, but how they will make society accept it by doing public relations. Here's a quote:

> Alongside a focus on influential opinion-makers, we also seek to build up a popular base of support among our natural constituency, namely, Christians.

— *The Wedge Strategy; Phase II: Publicity and Opinion-making*

So the Discovery Institute intends to defeat science by doing politics.

After the Leak

Of course, once the embarrassing *Wedge Strategy* became common knowledge, the Discovery Institute and its leaders tried to distance themselves from it. They now say that it isn't important. But they would say that, wouldn't they?

They know that the *Wedge* document is embarrassing to them. Here's what William Dembski, whose name appears several times in the Wedge document, had to say about it in 2002: "the wedge metaphor has even become a liability. To be sure, our critics will attempt to keep throwing the wedge metaphor (and especially the notorious wedge document) in our face."[2]

This doesn't sound like they're very proud of their work, does it?

However, though ID is probably the first movement to use a public relations campaign to try to "defeat" evolution, it is not the first movement to try to ignore evolution through political means. And politics, as we all have heard, makes for strange bedfellows. Do you suppose that the folks promoting ID know that the last major political campaign against evolution was done by Josef Stalin himself, in the old Soviet Union?

2. Dembski, "Becoming a Disciplined Science: Prospects, Pitfalls, and Reality Check for ID," http://designinference.com/documents/2002.10.27.Disciplined_Science.htm.

Chapter 8

Why Denying Evolution Can Get You into Trouble and Cause Mass Hunger, Too

Do you really want to be on the side of history that starved millions of people? Do you really want to agree with Josef Stalin?

The last major political group to oppose Darwin's theory of evolution on ideological grounds was the Communist Party of the Soviet Union under Josef Stalin. You won't find that in the ID literature, but you will find it out if you read about a man named Trofim Lysenko.

Josef Stalin, the murderous leader of the Soviet Union from the 1920s to the early 1950s, preferred Lysenko's ideas to the theory of evolution by natural selection. Lysenko was a biologist in the Soviet Union at the time. Lysenko claimed that acquired characteristics could be passed on from parents to offspring. This means, for instance, that if you dye your hair bright red every day, then your children will have bright red hair. It doesn't work that way, but Lysenko believed that it did.

Unfortunately, he wasn't just wrong. He was wrong and he had great political connections. He convinced Stalin that he was correct. Stalin wasn't a biologist, but he was a powerful dictator. As a result, Stalin announced that Lysenko was right, and anybody who disagreed with him was wrong, and he didn't believe any evidence to the contrary.

So Lysenko was put in charge of Soviet agriculture. He claimed that he had lots of scientific results, showing that wheat could be forced to acquire the characteristics that it needed to thrive in Russia's harsh climate within one generation. He predicted tremendous improvements in crop yields. This, of course, would have been a great boon to the Soviet people, who had recently experienced a terrible famine. Unfortunately, nobody could reproduce Lysenko's results, or get his ideas to work in actual farming. However, since Lysenko had succeeded politically, it didn't matter. He was proclaimed to be correct anyway.

Stalin Believed that Ideology Was More Important than Evidence— Just Like the Discovery Institute

One reason that Stalin liked Lysenko was because Stalin generally didn't like scientists, because scientists rely on observable facts rather than ideology. He kept physicists around because they could build bombs. But where biology was concerned, he got rid of the honest scientists and promoted Lysenko instead.

Lysenko's ideas agreed with Soviet ideology. Stalin believed that if people were simply good enough Communists, and if they really believed Communist ideals, then they could accomplish anything, even if they lacked food, clothing, and adequate materials for working. So the idea that wheat could be forced to perform at a newer and better level simply by exposing it to harsh conditions, fit right in with Soviet ideology.

Essentially, he promoted the idea that if the wheat were a good Communist, it would grow where it was told to, for the good of the nation, just as people under Communism were expected to do as they were told and perform miracles of productivity, even when they were given inadequate materials.

This argument also promoted the idea that people could be genetically changed into strong people and good Communists just by exposing them to the correct environmental factors. If wheat could be forced to obtain new genetic characteristics, then so could human beings, and a new Soviet man could come into being—one who was far superior to his Western counterparts, simply because he had been raised by good Communists.

Lysenko had a lot in common with ID proponents. First, he didn't do experiments whose results could be duplicated by anyone else. Second, he was not an "academic" scientist, and this appealed to the "anti-elitist" Soviet leaders. Third, he succeeded through politics, not through facts. This is very much how ID proponents wish to succeed. Lysenko's views agreed with Communist political ideals, and this is why they were promoted.

Unfortunately for the Soviets, there is a difference between political ideals and biological facts. Our Christian Right is under a similar delusion that they can dictate scientific facts. In fact, ID is just a form of "political correctness" for the Christian Right.

Figure 8.1 Lysenko with Khrushchev.
Lysenko (far left) with Nikita Khrushchev (center left), the Soviet leader. Clearly, Lysenko
was favored by Communist Party leadership.

Lysenko turned out to be a disaster for the Soviet people. Soviet agriculture, which was based on Lysenko's ideas, produced disastrous harvests. Lysenko's political power remained strong, however, under both Stalin and Stalin's successor Khruschev. Because of many years of disastrous harvests, the Soviet Union had to buy enormous amounts of wheat from the United States in the 1960s, 1970s, and 1980s just to feed its people.

Essentially, the Soviet Union and its agriculture never recovered from the anti-scientific mess that Lysenko left behind. However, now, the land of the old Soviet Union exports wheat.

I hope that anti-evolution and anti-scientific prejudice in the United States won't lead us into dependence on our enemies for basic survival, the way that the Soviet Union's anti-evolution political dogma led it to have to import American wheat, just for basic survival.

Chapter 9

What Science Is

WHAT IS SCIENCE?

Science is a method of investigation. That's all it is. However, this method of investigation has produced so much accurate, useful knowledge that societies using this knowledge have been able to live in a state of health and well-being that was unheard of in previous centuries.

Scientific investigation has a number of important components, both in how the research is done, and how it is reviewed.

Observation

Science starts with careful observation. Only things that are actually seen, actually heard, or can be detected by specific instruments can be used. The instruments themselves must be available for public scrutiny, so that all aspects of scientific observation are done in a transparent manner. Transparency is of the utmost importance, since results are not considered valid if they can't be reproduced by others.

Measurement and Quantification

You can't compare one thing to another without measuring them both. We figured out the orbits of the planets around the sun by doing numerous painstaking measurements of where the planets were in the sky, at what angle, in what places and at what times, for many years on end. But once we had figured out how the orbits were shaped and how much time they took, we were able to figure out exactly where the planets would be in the sky for years and even centuries to come. Astronomers today can tell you exactly where the planet Jupiter will be in relation to the earth, the sun, and to the other planets in our solar system in ten thousand years' time. Easily. In science, measuring things, and figuring out how to measure them (this is called quantification) is fundamental to what we do. Once some phenomenon has been measured carefully and repeatedly, then scientists can try to figure out what patterns there are, if any. The patterns can be used to make predictions about the results of future observations or future experiments. If these predictions have measurable outcomes, then experiments can be done or observations made, and their results can be assessed in a meaningful way.

Hypothesis and Prediction

A hypothesis is an idea that can lead to a prediction. To do an experiment and have solid experimental evidence, you need to make a specific, testable, quantifiable prediction that is based on your hypothesis.

Experiments and Controls

When you do an experiment, you are testing your prediction. To adequately test a prediction in this way, you must create the circumstances in which you can find out if the prediction came true, or not. This requires two careful procedures. The first is removing all other factors that might influence your results, so that you are testing only the one, single prediction. The second procedure is to do the entire experiment twice. The first time, you include the single factor that you are testing. The second time, you don't include it, so that you can see what your results would look like in the absence of the tested factor. That way, whatever difference you see between the first run and the second one can be reasonably attributed to the factor you were testing, and not just to the fact that an experiment was being done. This is called doing a controlled experiment, and it's the basis for a whole lot of science.

It is also sometimes necessary to repeat this whole process many times, just in order to reduce the probability that some of those outcomes were flukes that had nothing to do with the hypothesis that you were testing. A single experiment may take years to do.

This is why science often seems to take so long, and why it is so laborious. Day-to-day science is painstaking and goes at a snail's pace, but the small bits of reliable knowledge that have been gathered this way have accumulated to the point now where our knowledge about the material world is vastly superior to our knowledge at any other time in human history. Science is very slow, but very effective.

ID is not done this way. ID proponents simply refuse to make testable hypotheses about the material world, and they certainly do not test them. So without predictions, experiments, or quantifiable results, ID cannot realistically claim to be science.

It's not just me who says that ID doesn't make predictions. Here's William Dembski on the subject:

> Yes, intelligent design concedes predictability.[1]

Statistical Analysis and Conclusions

After an experiment has been done, you don't get to simply say "See! I was right!" You have to analyze your results using statistical tests, in order to make sure that your results didn't happen by chance. You also have to give the most cautious explanation possible for your results, and not make any claims that your evidence doesn't solidly support.

Publication in Peer-Reviewed Journals

This is another time-consuming process. After all the experimentation and analysis, you then have to write about your experiment, starting with the hypothesis and the prediction, and including everything about the methods you used and the results you got and the analyses you did. You then

---

1. Dembski, *Is Intelligent Design Testable? A Response to Eugenie Scott,* http://www.discovery.org/a/584.

need to send this written report to a journal that is run by other scientists. Those other scientists then get to pick apart your work, and figure out every possible way in which you could be wrong. If they don't like your report, they can refuse to print it. If they think it's okay, then they may print it in their journal, where it will be read by other scientists, and probably picked apart some more. Sending your report to a journal run by other scientists who know your field is called peer review. People can, of course, publish their results on the internet without any peer review at all, but most people won't believe results that are published without peer review, and most scientists certainly won't.

Reproducibility

Reproducibility is a key factor, and perhaps the biggest, most important test in all of science. Once your results have been published in a peer-reviewed journal, other scientists must be able to obtain your results, using the methods that you published in your paper. If repeated, honest attempts by more than one laboratory to reproduce your results fail, then your results will be considered to be invalid.

Many interesting ideas, hypotheses, and theories have been rejected because the results claimed by the original scientists couldn't be reproduced. For instance, some very exciting initial results that promised a new source of cheap energy through cold fusion were rejected because after repeated attempts, the authors' results could not be reproduced.

This was not a case of a theory or hypothesis being rejected because it was revolutionary, even though it was. Many people wished the results were true, because they could have solved the world's energy problems. But the results wound up being considered to be invalid because they could not be reproduced.

By contrast, ID papers do not even have a section of their papers where they describe their methods, so no one else would be able to verify their results. This, all by itself makes ID papers the kind of papers that would be rejected by any peer-reviewed scientific journal.

WHAT IS A THEORY?

A theory is a model of how the material world works. It is not just a single hypothesis, or even a bunch of hypotheses. It is a model for how some very large phenomenon works. It should actually generate testable hypotheses.

The germ theory of disease is a great example of a theory. The germ theory of disease simply states that microorganisms are the cause of many diseases. This theory has been extremely helpful to the human race. By understanding that microorganisms are a source of disease, and not some supernatural power, the human race has been able to create safe drinking water, understand safe food-handling practices, and create vaccinations and antibiotics that prevent and cure many diseases. None of these things were possible before the age of modern science.

Interestingly, germ theory doesn't explain all disease. For instance, cancer is not caused by microorganisms. However, understanding of germ theory combined with not accepting supernatural explanations has allowed us to understand what cancer is, and we are now able to successfully treat many forms of it.

No Supernatural Explanations

One aspect of science that is crucial is that supernatural explanations for the material world are not used. This has proven very useful to science, since it means that scientists keep investigating things, and have been able to explain many things that were previously unexplainable. If we had stopped with an explanation like "God did it," we would never have cured polio or smallpox, diseases that were once so common and so deadly that people died in the thousands from them; now, many Americans have barely heard of them.

It is not the case that paranormal explanations are rejected out of hand. They are rejected because they don't provide solid evidence for themselves. Scientists over the years have been willing to investigate anything, including the paranormal. Unfortunately, every time that a scientist does a controlled experiment on people claiming to have paranormal abilities, it turns out that they don't. Scientists still haven't found anything in a controlled experimental setting that supports claims of anything supernatural.

SCIENCE AND THE MATERIAL WORLD

By definition, science doesn't know everything. If we did, nobody would do any more scientific research.

But as a means of finding out about the material world, it has an unsurpassed track record.

In the past 400 years since the invention of modern science, the knowledge gained through this method of investigation has improved the health, safety, longevity, and well-being of humankind far beyond anything that our ancestors could have dreamed of.

Chapter 10

Why Accepting Science and Evolution Can Lead to Better Values for All Humankind

The facts that scientists have obtained through the scientific method have not always been used wisely or well. However, this is not the fault of the means that we used to obtain the facts. Just as we do not blame the invention of the wheel for human deaths, even though using wheels has meant the deaths of countless human beings through automobile accidents, we should not blame science when human beings make poor use of the knowledge that they have.

WHY RELIGIOUS PEOPLE ARE SOMETIMES DISTURBED BY EVOLUTION

Two of the many things that religions do are first to explain the world, and then to present a moral code. They appear to derive their authority to present the moral code from the fact that they explain the world, or so it seems. Often, the explanation of the world—the creation myth—foreshadows the moral code.

So people sometimes believe that if you replace the creation myth with a scientific explanation of the world, then there will be no moral code. People are often attracted to ID because they want to have a Creator—that is, a God—because that way they can still insist on a moral code that they believe was handed down by God.

In fact, explaining the world and having a moral code are unrelated.

Science is really, really good at explaining the material world. As a method of investigating the material world, it can't be beaten. In the past 400 years, it has advanced our knowledge of the physical world far more than in all the rest of human history put together. This has lead to better human health and increased our life-spans and opportunities for happiness and spiritual enlightenment and growth far beyond what most of our ancestors ever experienced.

But, science is a method of investigating the material world. Period. That's all it is. It is silent on the subjects of values and morality. Those are different subjects.

We need to accept that explaining the material world and putting forward a moral code are separate jobs for the human race to do, and we need to start doing a whole lot of talking about values, without invoking anybody's God or creation myth.

MORAL CODE, OR A VALUES FREE-FOR-ALL?

What does it mean to be good? People want a good, solid moral code. Most of us believe that the moral code that everyone else in the US and the world should follow is the one that we ourselves believe. However, getting everyone else to fall in line behind that idea just doesn't work. So we need to have a long, national conversation about values, including the ones we want to have, and the ones we want to reject.

I know what you're going to say. People want a simple moral code that they can follow, and they don't want to have a national free-for-all about values. I understand. But the problem is that there is a national free-for-all about values going on right now whether any of us likes it or not. It is our duty to speak up, since I guarantee you that other people with different values will speak up, and often they will be very well funded and very well organized. So if we don't speak up, then other people's values may well be the ones adopted by the majority of the US and the world.

We often think of morality as being changeless, but some of this is an illusion. Slavery was once thought to be a perfectly reasonable and acceptable part of a good moral code. Today, you would find very few people in the United States who believe that slavery is good, despite the fact that many people prior to the Civil War pointed to the Bible for evidence that slavery was acceptable.

Religion has a major role to play in the development of values. Beauty, justice, and kindness, for instance, are not addressed directly by science. Religions can and should address these issues, and derive their moral authority from the good that can be done in the world by adhering to the ethics and moral code that they propose. Science can make an important contribution to this discussion by pointing out facts as it discovers them. Without these discussions, the world is left rudderless. We also need to set up institutions that help lead to good human behavior.

It should be presumed that if religions either opt out of a discussion of morals because they believe that their moral authority has been undercut by science, or if they ignore the facts that are presented by science, then others less caring than they are will happily rush in and provide the guidance that is otherwise not being given. This is what happens far too often in the present world, and it shows an open field in which science and religion can both participate, and may even act as partners.

Chapter 11

Bad Design—the Birth Canal

To return the subject of bad design, here is a picture of a baby crowning. This is the part of childbirth where a baby's head has to fit through a circle of bone that is smaller than the head is.

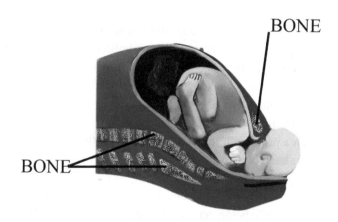

Figure 11.1   A baby crowning.

I've labeled the bony parts of the woman's pelvis. The woman's pelvis makes a circle of bone around the birth canal, and a baby being born has to fit its head through this circle. It isn't a good fit.

Here's a picture of the whole birthing sequence.

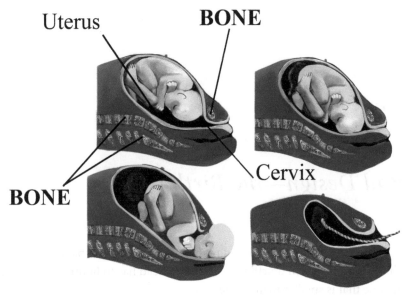

Figure 11.2   The birth process.

As before, I've labeled the bone. You can see how the baby's head has to be squeezed pretty hard to fit through this bottleneck. As we all know, in the old days this frequently killed the woman, the baby, or both. These days, we often avoid the birth canal altogether by performing Caesarian sections. These are being done in ever-increasing numbers.

One would think that a benevolent Creator would not make childbirth into such a problem in the first place. In fact there are simple things that could have been done better, if only we *had* been designed rather than evolved.

Let me explain. The main problem from a design standpoint is that we walk upright while being very smart. These two attributes have opposing requirements. Walking upright favors people with narrow hips, which make walking much easier and more efficient. Being smart, on the other hand, requires large heads. Large heads require large birth canals. Large birth canals require *wide* hips.

Now, a wise Creator could have solved this engineering problem easily, by doing something like—this!

Figure 11.3   Kangaroo and joey.

Look! The baby develops *outside the body* of its bipedal mother, in a nice comfy pouch complete with a nipple for nursing.

Animals like kangaroos give birth to very small, embryo-like young that are placed in a pocket on the outside of the mother's body. This is where they continue their development. That's the way to do it if you're going to be a biped.

Of course, another way to do it would have been to give us four feet as well as two hands. This would have placed less weight and stress on the hips, allowing the pelvis and birth canal to be wider without sacrificing the ability to walk well. Why didn't the Creator do that?

It's simple. We evolved, rather than being designed. Women's hips are narrow enough that they can walk, because any woman who couldn't walk would die before she could reproduce, in the natural environment. Most women's hips are wide enough, on the other hand, that children can be born . . . most of the time. The end result is an uneasy compromise that doesn't work very well, and is very hard on some individuals. This was clearly not done by any intelligent Creator. In fact, if this is the best that the Creator can do, then the Creator has a lot of explaining to do.

The ID lobby has a lot of explaining to do, too. They blithely dismiss the deaths of millions of women and children. In his book *The Design Revolution*, Dr. Dembski says, "It would be nice to have all the functionality of a female pelvis along with easier delivery of children . . . But when the suboptimality objection is raised, invariably one finds only additional functionalities mentioned but no details about how they might be implemented."[1]

On the contrary. I have shown that the kangaroo method of bearing children would be very well suited to the challenge presented to human females, with humans' combination of large heads and two-legged locomotion. Dr. Dembski should study zoology more carefully. There are millions of kangaroos in the world, yet somehow Dr. Dembski missed out on seeing a picture of even one of them.

As to Dr. Dembski's statement that it would be "nice" if women could bear children more easily—I have one word for him: fistula.

Now Dr. Dembski, being a man and a theologian rather than being either a woman or a biological scientist, may not be aware of obstetric fistula, so I will explain it. *Obstetric fistula* is a medical condition in which a passage forms between a woman's birth canal and her urinary bladder. Or between her birth canal and her rectum. When she has fistula, the woman dribbles either urine or feces, constantly, for the rest of her life.

It occurs when there has been a difficult, obstructed labor. This happens frequently. If a birth takes a long time, which it often does, then the baby's head can press against the mother's pelvis and cut off the blood supply to those delicate tissues. When this happens, the tissue dies. The dead tissue falls away, and the woman is left with a hole that forms a connection between her birth canal and either her bladder or her rectum. It heals that way, leaving the woman with a permanent, uncontrollable connection between her vagina and one of those two places.

When this happens, she is often abandoned by her husband and family, and ostracized by society because she is always filthy and foul smelling and dribbling. She can't hold a job, and frequently isn't allowed to prepare food. She can't have any more children. These women frequently die from urinary tract infections or infections of the vagina caused by feces.

---

1. Dembski, *The Design Revolution: Answering the Toughest Questions About Intelligent Design* (Downers Grove, IL: InterVarsity, 2004), 60.

Is this any way for the Creator to act? Remember, childbearing is a normal process. Women get pregnant all the time. It's not like this is an unusual thing to have happen.

It is estimated that 5 percent of all pregnant women will have an obstructed labor. In places with modern medical care, women can get emergency obstetric care such as C-sections. In these places, obstetric fistulas have been largely eliminated. But in parts of the world where women do not have ready access to modern medical care, approximately two million women have untreated obstetric fistulas, with 100,000 otherwise healthy women developing them every year.

For example, in Ethiopia, there are approximately 100,000 women who suffer with untreated fistulas right now, and another 9,000 women who develop them every year.[2]

If this is Dr. Dembski's idea of intelligent design, I shudder to think of what his idea of lousy design is.

Here is a picture of a healthy female reproductive system.

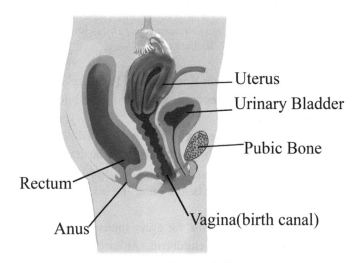

Figure 11.4   A healthy female human reproductive system.

2. The Fistula Foundation, http://www.fistulafoundation.org/aboutfistula/faqs.html.

Here is a picture of a fistula between the vagina and the rectum.

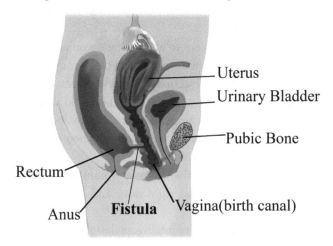

Uterus

Urinary Bladder

Pubic Bone

Rectum

Anus   **Fistula**   Vagina(birth canal)

Figure 11.5   A female human reproductive system with a fistula.

An obstetric fistula is a passage that forms between the birth canal and either the rectum or the urinary bladder. It is the result of a difficult labor in which some of the vaginal tissue is killed in the process of giving birth. This illustration shows a fistula between the rectum and the birth canal

Of course, Dr. Dembski only thinks it would be "nice" if women didn't have this problem. Or all the other ones associated with childbearing, problems like life and death.

Did Dr. Dembski not notice that giving birth is actually life-threatening? Has he checked mortality records for women giving birth in the ages before modern, scientific medicine? Women and infants died in childbirth by the millions. Women wrote their wills before going into labor, because they knew they might not survive it.

In England in the early 1700s, for every thousand live births, ten to eleven women died as a result of childbirth.[3] At the beginning of the twentieth century in the United States, six to nine women died for every 1,000 live births.[4] Even today, in areas where modern medicine is not available, women have a frighteningly high possibility of dying during childbirth. In Afghanistan's province of Badakshan, fifty to eighty women died for every

3. Geoffrey Chamberlain, "British maternal mortality in the 19th and early 20th centuries," *Journal of the Royal Society of Medicine*, November, 2006, 99 (11), 559–63.

4. Centers for Disease Control, "Achievements in Public Health, 1900–1999: Healthier Mothers and Babies," *Morbidity and Mortality Weekly Report*, October 1, 1999, 48 (38), 849–58.

thousand live births in 2002![5] Given that women tended to have numerous children in the old days, it was very common for a woman to not live through bearing all her children. Even in 2010, a woman in Afghanistan had a one in thirty-four chance of dying as a result of pregnancy during the course of her lifetime.[6] In areas where modern medicine is not available, babies also have a significant chance of dying just from the process of being born. In rural Bangladesh in 2011, seventeen out of every thousand babies died as a result of the birthing process.[7]

Perhaps Dr. Dembski just doesn't see millions of women dying as a problem. But I say that an easier arrangement for childbearing would not have been merely "nice." *It would have saved countless millions of women's lives.* But I guess that doesn't matter to Dr. Dembski.

Then there is the final malfunction of the human reproductive system: the problem of spontaneous abortion.

THE DESIGNER LOVES MISCARRIAGES

Many babies never even make it *into* the birth canal alive. Many fertilized eggs do not go on to become live babies. In fact, *over 31 percent of all fertilized eggs do not become live babies.*[8,9] This is a conservative number.

Think about it. Nearly one-third of all pregnancies spontaneously fail. This is a cruelty to all the women who have ever endured the agony of a miscarriage. And ID proponents who pretend that our systems are well designed only add to this cruelty by making women think that it must be their fault if a pregnancy spontaneously miscarries.

5. Linda Bartlett, Sara Whitehead, Chadd Crouse, Sonya Bowens (US Centers for Disease Control and Prevention) and Shairose Mawji, Denisa Ionete, Peter Salama (UNICEF), "Maternal Mortality in Afghanistan: Magnitude, Causes, Risk Factors and Preventability, Summary Findings," November 6, 2002, Joint Press Release.

6. Matthew Ellis, Kishwar Azad, Biplob Banerjee, Sanjit Kumer Shaha, Audrey Prost, Arati Roselyn Rego, Shampa Barua, Anthony Costello, and Sarah Barnett, "Lifetime risk of maternal death (1 in: rate varies by country)," http://data.worldbank.org/indicator/ SH.MMR.RISK.

7. "Intrapartum-Related Stillbirths and Neonatal Deaths in Rural Bangladesh: A Prospective, Community-Based Cohort Study," *Pediatrics*, vol. 127 no. 5, May 1, 2011, e1182 -e1190 (doi: 10.1542/peds.2010–0842).

8. A. J. Wilcox et al., "Time of Implantation of the Conceptus and Loss of Pregnancy," *New England Journal of Medicine*, 1999 (340), 1796–99.

9. A. J. Wilcox et al., "Incidence of early loss of pregnancy," *New England Journal of Medicine*, 1998, 319 (4), 189–94.

What's more, *with a 31 percent spontaneous abortion rate, the Designer is the world's biggest abortionist!*

Now, I'll stop talking dirty to you for a while, and I'll talk about my handbook instead.

Chapter 12

The Handy-Dandy Intelligent Design Refuter, Part 1

Back when creationism was first being used to try and take over American education, the creationists published a handbook called *The Handy Dandy Evolution Refuter*. It was very popular with creationists, but both scientists and the courts agreed that it didn't refute evolution very well at all.

However, the name is great, so I'm using it. Below, I have listed many of ID's favorite sayings and arguing methods, and I show you how they are wrong. So this is my Handy-Dandy Intelligent Design Refuter.

"TEACH THE CONTROVERSY"

One of the ID lobby's favorite ploys is to say that we should "teach the controversy" about evolution and ID. They argue that if students are told about both "sides" of the argument, then they can choose for themselves, just the way students are allowed to choose what side they believe when we tell them that George Washington was the first president of the United States, or the way students are allowed to choose whether or not they believe that water is formed by two atoms of hydrogen and one atom of oxygen.

In other words, there actually is no controversy. It is like asking a geologist to teach the "flat earth" controversy. Again, there is no controversy, and a few people with crackpot ideas can't create one just because they want to.

"But wait!" A flat-earth proponent would cry. "There are things that the spherical-earth theory doesn't explain! It doesn't explain the Grand Canyon or Mount Everest. Something that is spherical can't have these huge irregularities. What's more, there are other problems with the spherical-earth theory. Even spherical earthists admit that the earth bulges out at the equator. Since there are problems with the spherical-earth theory, obviously then the flat-earth theory is just as correct as the spherical-earth theory. Both of these should be taught in schools as being equally possible, and we should point out that there is disagreement among scientists as to the exact shape of the earth."

Doesn't this sound like ID promoters?

Please keep in mind that compromise is not a good solution. If you tried to compromise between the incorrect idea that the earth is flat, and the fact that it is a sphere, you would wind up with people believing that the earth is shaped like a throat lozenge. It makes no sense. What's more, navigation, satellite signaling, and many other useful skills and technologies would suddenly have no foundation, even though they work. I pity any passenger on an international jet liner whose pilot truly believes that the earth is flat. It could end tragically.

As a result of the manufactured controversy about evolution, sometimes crackpots insist on putting stickers on biology textbooks, disclaiming evolution. For instance, here is the sticker from Cobb County, Georgia.

This textbook contains material on evolution. Evolution is a theory, not a fact, regarding the origin of living things. This material should be approached with an open mind, studied carefully, and critically considered.

Approved by
Cobb County Board of Education
Thursday, March 28, 2002

Figure 12.1   Text of the original sticker made by the Cobb County, Georgia Board of Education.

If we apply their logic to a spherical earth, since there are some who claim that the earth is flat, here is what a "spherical-earth controversy" sticker would look like.

This textbook contains material on the spherical earth. The shape of the earth is a theory, not a fact, and many people do not accept this theory. This material should be approached with an open mind, studied carefully, and critically considered.

Approved by
Your Local School Board

Figure 12.2   An imagined "spherical-earth controversy" sticker.
Reproduction of the first sticker with new text.

The weight of scientific evidence is entirely in favor of evolution, just as it is in favor of a spherical earth. Students are not taught that both the flat-earth and spherical-earth "sides" have their merits, and that they are free to choose between them. This would be a disaster for the students' understanding of the world and preparation for life in it.

Interestingly, ID promoters do not advocate teaching any other controversies, or allowing students the "freedom" to make up other facts as they go along. The only "controversy" that ID is willing to brook is the one that fits in with their religious beliefs, yet they claim that theirs is not a religious argument.

I say that if this "controversy" should be taught, then students should be allowed to make up other controversies as well. For instance, why should pi have the value of 3.1415926 . . .? Why not 3.1? This would be much easier for calculations. I personally would prefer that French grammar be taught as a controversy, so that I wouldn't have to learn all those difficult rules. My grades in French would have been immeasurably better if I had been allowed this boon. Physics would be much easier if we didn't need to memorize all those equations, or if we could argue that there was controversy surrounding which ones to use and when. After all, there are still some arguments among physicists about various aspects of physics, so therefore, if evolution should be taught as a "controversy" because there is still research being done on it, so not all facts are known, then the same is true of physics. Gravity will never be the same again.

"ONLY A THEORY"

This brings up another of ID's favorite sayings. That is that evolution by natural selection is "only a theory."

Gravity is only a theory. It is explained in Einstein's *theory* of general relativity. I invite ID promoters to walk off a cliff and see if something's being a theory makes it untrue.

The germ theory of disease is also a theory.

Diseases used to be thought to be caused by god, usually as punishment, or to test us; mental illness was sometimes thought to be possession by demons. This is why people died so young in the days before science.

Fortunately, scientists and medical doctors slowly figured out, through both observation and testing, that many diseases are caused by microorganisms, which they called germs. Understanding and accepting the germ theory opened up vast new avenues for preventing and treating disease. Vaccination, antibiotics, and disease prevention though clean water would all be impossible if we didn't understand the germ theory of disease.

Unfortunately, there was opposition to it. For instance, religious arguments were used to oppose people's being inoculated against smallpox, because smallpox was believed to be "a judgment of God on the sins of the people," and that "to avert it is but to provoke him more." People said that inoculation was "an encroachment on the prerogatives of Jehovah, whose right it is to wound and smite."[1] Arguments like these were used to oppose lots of breakthroughs that came about because of germ theory.

However, germ theory brought results. People were much healthier in measurable, testable, dramatic ways when germ theory was applied.

1. A. D. White, *A History of the Warfare of Science with Theology in Christendom*, http://vserver1.cscs.lsa.umich.edu/~crshalizi/White.

Germ Theory Was Accepted, Even Though It Doesn't Explain All Disease

Not all diseases are explained by germ theory. For instance, cancer and diseases caused by vitamin and mineral deficiencies are not explained by germ theory. Fortunately, sensible people did not go backward, and then attribute these ailments to God, but did research and discovered what non-supernatural causes produced these diseases.

We Are Lucky to Have Been Born After Germ Theory Was Accepted

We are very, very lucky that our ancestors accepted the evidence for germ theory, since this has allowed us, their descendents, to enjoy much healthier lives than they did.

What's more, our ancestors didn't just accept germ theory, they followed up on its implications, even though it was only a "theory." They built sewage treatment plants that allow us to have clean, germ-free water. They learned about vaccination, which couldn't have been achieved without accepting and acting on the germ theory of disease. We now have antibiotics, vaccinations, clean food, and clean water, all as a result of our ancestors' having accepted the germ theory of disease, even though there was religious resistance to it. I invite those who don't like theories to drink contaminated water, after they have walked off the cliff. Even if they survive the fall, they won't survive the bad water.

Here is a partial list of diseases that we can now cure or prevent because we accepted the germ theory of disease:

DISEASES PREVENTED BY VACCINES: cervical cancer, diphtheria, haemophilus influenzae type b, hepatitis A, hepatitis B, human papillomavirus, influenza, Japanese encephalitis, measles, mumps, pertussis (whooping cough), pneumococcal disease, polio, rabies, rotavirus, rubella (German measles), shingles, tetanus (lockjaw), varicella (chickenpox), variola (smallpox), and yellow fever.[2]

Smallpox, which used to kill millions of people, was eradicated in 1979, due to worldwide vaccination campaigns.

2. List from U.S. Center for Disease Control, http://www.cdc.gov/vaccines/vpd-vac.

Bacterial diseases that can be cured or treated using antibiotics: anthrax, Bubonic Plague, staphylococcus (Staph) infections, streptococcus infections (such as strep throat), pneumonia, gonorrhea, meningitis, salmonella, cholera, tuberculosis, leprosy (Hansen's Disease), syphilis, tuberculosis, typhus, and Lyme disease.[3]

Evolution as Fact and Theory

What's more, evolution itself, that is, species changing over time, is an observed fact. The only part that is a theory is the mechanism, which is called natural selection. Evolution as Charles Darwin proposed it has the full title of "the theory of evolution by natural selection." This is because there were other theories of evolution that preceded natural selection, such as one proposed by a scientist named Lamarck. But Darwin's theory of natural selection had all the evidence, so eventually, after much testing, scientists accepted it.

"THERE'S THIS THING THAT EVOLUTION HASN'T EXPLAINED—DOESN'T THAT MEAN THAT INTELLIGENT DESIGN IS CORRECT?"

ID promoters are constantly searching for things that they can claim are not explained by evolution. These days, that usually means some small, microscopic part of an organism that they say must have been created, since they personally can't think of a way for it to have evolved.

This is like saying that the theory of gravity is wrong because you personally can't explain why helium balloons float. Helium balloons seem to defy gravity, but they are easily explained by science. However, you have to be willing to do scientific research to find out why, and you have to be willing to listen to people who do scientific research on that subject. ID promoters are famous for ignoring any research that doesn't fit their claims, no matter how many times they are publicly told about it.

---

3. See Lansing M. Prescott, John P. Harley, and Donald A. Klein, *Microbiology*, 5th ed. (New York: McGraw-Hill, 2002), chapter 39.

God of the Gaps

Using a Designer as a catch-all for things you don't understand is also a rather sad sort of religious argument. It essentially says that God (or the Designer) is nothing more than a filler-inner who will be used to explain a gap in our knowledge, until science explains it. After science explains that particular gap in our knowledge, then God will be relegated to even smaller gaps, and then smaller ones still. This is known as the "God of the gaps" argument for the existence of God. It is a rather sad way to form a philosophy.

"SCIENTISTS HAVEN'T AGREED ON ALL THE DETAILS ABOUT EVOLUTION—DOESN'T THAT MEAN THAT INTELLIGENT DESIGN IS CORRECT?"

This is another ploy that ID tries. Scientists are always doing research on unanswered questions. That's what science is. If all questions were answered, science would no longer be needed. When scientists do experiments or make observations, they frequently answer the question they were working on, but other questions pop up as a result of this work. Some fundamental questions go unanswered for a long time. For instance, modern physicists can't figure out if light, at its most fundamental level, is a particle, a wave, or both—or exactly what light *is* in an absolute sense. This does not cause us to throw away all of modern physics and accept a religious explanation of light. This is a good thing, too, since there's nothing in the Bible about lasers.

It's the same in biology. All the evidence in biology supports evolution by natural selection, but questions about specific organisms or questions about the rate of change during evolution always occur. This doesn't cause us to throw away all of modern biology—which is based on evolution—and accept a religious explanation of biology such as ID.

"MICROEVOLUTION HAPPENS, BUT MACROEVOLUTION DOESN'T"

One of ID's favorite ploys is to claim that they accept microevolution but not macroevolution.

When biologists talk about microevolution, they mean changes in the gene pool of a population over time that are relatively small, so the changes wouldn't result in the newer organisms being considered a different species.

To biologists, macroevolution means lots of small changes in the gene pool of a population over time. Eventually, these accumulate to the point where the population is considered to be a new species. In other words, the new organisms wouldn't be able to mate with their ancestors, even if we were able to bring them together.

When biologists use these different terms, it is for descriptive purposes only. When ID promoters use them, however, they believe that they are describing two supernaturally different processes. The essence of what constitutes microevolution is, for IDers, different from the essence of what constitutes macroevolution. ID promoters behave as though there is some magic line between microevolution and macroevolution, but no such line exists as far as reality is concerned. Macroevolution is merely the result of a lot of microevolution over a long period of time. So the difference between macro- and microevolution is this: one's bigger.

For example, ID promoters claim that genetic changes such as antibiotic resistance only count as microevolution, while genetic changes that eventually result in new species count as macroevolution. One common way to put it is to say that dogs may change to become bigger or smaller, but they never become cats. So they say that microevolution may occur within the dog species, but macroevolution never will. Of course, they don't explain how dogs are related to their relatives such as wolves and hyenas, or their extinct relatives like dire wolves. ID promoters just like to draw lines where none really exist.

The Discovery Institute's Argument that Microevolution and Macroevolution Are Magically Different—Like the Tobacco Research Institute's Argument that Cigarettes and Cigars Are Magically Different.

The microevolution argument is a tactic that could have been taken straight from the tobacco lobby. At one time, the tobacco lobby admitted that cigarette smoking was linked with cancer. *But*, they said, it was not proven that cigars were. It was a distinction without a difference, just as ID is trying to do with microevolution and macroevolution. The difference? One's bigger.

The same is true for microevolution and macroevolution. The only difference is size.

As usual, men think that size matters.

THE EVOLUTIONARY SPECTRUM

ID claims that there's a magic difference between a few genetic changes (what they call microevolution) and many genetic changes (what they call macroevolution). This is like claiming that there's a magic difference between the light waves that we can see, and X-rays, which we can't see, but which we know exist because we've discovered them through science. The only real difference between light waves and X-rays is their wavelength. In other words, one's bigger.

Here's how it works. There are a whole bunch of waves involving light particles that travel through space. This is known as the electromagnetic spectrum. The only difference between these waves is their length. Other than that, they have the same composition.

Modern physics has not nailed down every aspect of what electromagnetic waves truly *are*. But we have found, measured, and used these waves in many different ways; so we know that they exist and we know a lot about what they do. That goes for both the electromagnetic waves we can see, which are visible light, and those like radio waves, X-rays, and gamma rays, that we can't see, but we know what their effects are.

Here are the waves, and their different lengths: we start with radio waves, which can be very long (ranging from many miles down to 1 millimeter), to infrared in which the waves are shorter (ranging from 1 mm to 7/10,000ths of a millimeter [that is, 0.7 microns]), to visible light (from 0.7–0.4 microns), to ultraviolet light (400 nanometers [0.4 microns]–10 nanometers) to X-rays (10 nanometers–.01 nanometers), to gamma rays (.01 nanometers or smaller). All these different wavelengths are a part of the electromagnetic spectrum. They are the same things, just different wavelengths. A few we can see, most we cannot.

It's the same with evolution. We have found it, measured it, used it, and made predictions from it, so we know that it exists, both when we can see it happening at the microevolutionary end of the scale, and when we can only see its effects, at the macroevolutionary end of the scale.

INTELLIGENT DESIGN HAS NO PLACE IN THE TWENTY-FIRST CENTURY

It is an embarrassment to the United States that evolution by natural selection is still not accepted by a large fraction of the population, including some school boards. That it is even in question in the United States in the twenty-first century is shameful. Evolution is a nineteenth-century concept that has been repeatedly proven to be correct over the last 150 years, and it is ridiculous that is still hasn't been accepted by the population of the world's leading industrial and technological nation. We will not maintain our scientific and technological leadership if we continue to pander to science-deniers who hide behind religion.

For more handy-dandy ways to refute ID's favorite sayings, see Chapter 16.

Chapter 13

Intelligent Design Requires a Leap of Faith

So ID promoters freely admit that a population can undergo genetic changes. They even admit that small amounts of evolution do take place, the kind that results in small modifications to existing organisms. Remember, they call this "microevolution." So, for instance, they say that you might get dogs of different sizes, or horses of different colors, but you will not see dogs turning into a new species.

What they ignore is that if there are enough changes to a population of dogs, or any other organism, over time, then eventually that population will be considered a new species. If the new, genetically changed group of dogs could not interbreed with the dogs of the original population, then it would be considered a new species. That's the biological definition.

The key words in the preceding paragraph are the words *"over time."* For a vertebrate population to show these levels of changes, many generations over thousands of years would have to be involved. Since this is outside the span of a human lifetime, no single human scientist can chart the complete progress of a living vertebrate population as it becomes a new species, or even several new species.

However, new microorganisms evolve into being on a regular basis. For instance, the AIDS virus (HIV), other viruses, and bacteria, have short enough generations that human beings actually can observe the evolution of new organisms.

So what does ID do when confronted with the fact that microorganisms evolve into new species all the time, and that this is well documented by science? They declare that this evolution is "microevolution," and that means that, according to them, it doesn't count.

Likewise, when moths become a different color after the trees in their environment change color, ID promoters tell us that this doesn't count as real evolution but only as microevolution as well.

So, in other words, *if a change can be observed and measured, then ID promoters insist that it isn't real evolution.*

This proves that ID is religion, and not science.

Why? Because science is about measuring and testing. To be science, an idea must be measurable and testable. That is, you have to be able to make a prediction about the natural world, and be able to test that prediction. Without this testing against reality, an idea may be philosophy or religion, but it isn't science.

So by cleverly claiming that any change that can be measured doesn't count, ID promoters have made ID untestable.

By ID promoters' own definition, ID cannot be tested, since any experiment we could do to either prove it or disprove it would automatically be classified by ID promoters as being microevolutionary in nature.

This means that ID's promoters are stuck with an idea that cannot be tested scientifically.

ID proponent Michael Behe concedes, "You can't prove intelligent design by experiment."[1]

That means that ID is not science. ID's own promoters have defined it as being outside the realm of science.

Therefore, accepting ID requires, and must be taken on, a *leap of faith.* This means that ID can't be taught in science classrooms, no matter what its promoters say.

Now let's talk about some animals that ID proponents don't want you to know about.

---

1. Quoted in Claudia Wallis, "The Evolution Wars," *Time*, August 7, 2005, 32.

Chapter 14

Animals That Shouldn't Exist According to Intelligent Design

One of the most entertaining things about evolution is the weirdness of its results. Animals that no rational Creator would have come up with exist perfectly well in our evolved world because they work well enough, and survive from generation to generation.

Two of my favorite animals are ones that make no sense at all from a planning point of view, but are wonderful examples of the strange ways in which animals evolve.

WALKING FISH

Let's start with fish. Any decent ID proponent can tell you that fish were designed to live in water. They have gills to extract oxygen out of the water. They have fins and tails that allow them to swim. The idea that a fish might walk on land seems so obviously wrong to anti-evolutionists that they routinely ridicule the very idea that a fish might walk on land, and use it as an example for why they say that evolution is wrong.

Here is a photograph of a walking fish:

Figure 14.1   A mudskipper climbing onto a rock.

This is a mudskipper. It's a type of fish.

Mudskippers live in intertidal swamps and mud flats where water levels change every time the tide changes.

Figure 14.2   A mudskipper taking a walk on the beach.

When the tide comes in, this fish does what any sensible fish does when confronted with more water. It hops out and climbs a tree.

Now any decent design proponent can tell you that fish do not climb trees. But the mudskippers failed to read all those ID books and went right ahead and evolved in intertidal areas where getting out of the water to avoid predators—even by scrambling into trees—could help an animal to survive.

Here is a photograph of a mudskipper climbing a tree.

Figure 14.3   A mudskipper climbing a tree.

Of course, a decent design proponent could also tell you that if you are going to make climbing gear, you should not start with fish fins. Unfortunately for them, they didn't tell this to the fish. Mudskippers have special pelvic fins that help them climb trees.

Here is a picture of the fins that mudskippers use to climb trees.

Figure 14.4   A mudskipper's special fused pelvic fin that helps with climbing.

As climbing gear they don't work brilliantly, but they work well enough, and they help the mudskippers survive to reproduce.

It gets weirder. Not only do these mudskipper fish walk, skip, hunt, and climb on land, they even build their nests on land. Here is a photograph of a mudskipper's nesting burrow. On land.

Figure 14.5 A mudskipper's nesting burrow.
You can see the two eyes peeking out of the hole in the sand.

It gets weirder still. The eggs that hatch inside the burrow develop in air, not water. That's right, these fish breed out of water and develop their eggs out of water. What's more, when oxygen levels get low inside the burrow, the male gulps air (yes, air) from outside the burrow and transports it into the egg chamber in its mouth, and then releases the higher-oxygen air into the egg chamber, so the developing eggs get enough oxygen—*from the air.*

These fish have gills, but they do their breathing through their skins, and through sacs of air that they trap in their skins. The air sacs are basically simple lungs.

These fish can swim, and do, but they spend as much as 90 percent of their lives out of the water.

As usual, design proponents have failed to look at biology. As examples of design, these fish fail miserably, unless you posit that the Designer has a very weird sense of humor. But as examples of the strange and wonderful variety of life that comes out of evolution, these fish are marvelous.

THE IMMORTAL JELLYFISH

Now let's move on to one of life's most unfair arrangements. Human beings have always yearned for immortality. For thousands of years, our desire for immortality has been the driving force behind stories, legends, quests, and epics. Explorers and alchemists both strove to find it.

Major religions have been built around it. Worshiping one supernatural being or another has been said to be our means of achieving it. Yet somehow the Designer—a supernatural being—left us out when he really did make a living being immortal.

Instead, the Designer awarded immortality, this most sought-after state of being to . . . a jellyfish.

Here it is, the true immortal, known as the immortal jellyfish.[1]

Figure 14.6   *Turritopsis nutricula,* **the immortal jellyfish.**

This species of jellyfish, *Turritopsis nutricula,* can live, basically, forever. They can avoid death altogether. If these creatures are stressed by wounds, heat, starvation, or even simple old age, then they can simply revert to a juvenile form, make new baby copies of themselves, and start all over again (a diagram of the repeating life cycle of *Turritopsis nutricula* can be found in Appendix 3). This would be like an insect reverting to being a larva, and then making new copies of that larva, all of which can then grow up. If this were a human being, it would mean reverting to being a

1. Piraino, S., Boero, F., Aeschbach, B., and Schmid, V. Reversing the life cycle: Medusae transforming into polyps and cell transdifferentiation in *Turritopsis nutricula* (Cnidaria, Hydrozoa). *Biol. Bull.* 90:302–312.

child, making dozens of copies of that child, and then letting them all grow up. If the child, or the adult it becomes, at any stage in its life finds itself wounded, without food, senile, or otherwise in distress, it can again regress to childhood, re-multiply, and start all over again. Over and over again, without end.

So the immortal jellyfish doesn't just have eternal life . . . it has eternal youth. Or at least, continuously restarting youth. Ponce de Leon can eat his heart out! But seriously, why didn't the Designer do that for us?

Just think, we wouldn't need plastic surgeons anymore. And no more Botox! Think of all the charlatans we could do away with. No more megavitamin regimens! No more hormone replacement therapy!

And no more religions that tell you that you can have eternal life after you die but only if you do everything just right and nobody can tell you ahead of time if you're going to make the cut, and you don't get to see for yourself if it really works anyway because nobody ever comes back to tell you what it's like.

No wonder ID hates biology.

Chapter 15

Bad Design—the Human Throat

Here is a picture of the inside of the human throat. This is another example of really *bad design*. Our air passages and food passages meet and mix, sometimes with fatal results.

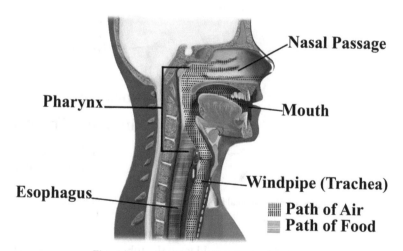

Figure 15.1   Human esophagus and trachea.

The pathway for air is marked with dots. The pathway for food and liquid is marked with lines. *Note:* these paths overlap in the mouth, and cross at the pharynx.

I have shown the parts where air gets inhaled with dots. I have shown the parts where food gets ingested with horizontal lines. The places where both food and air go are at the mouth, and farther down at the pharynx.

Most of the time, the air we inhale passes through the pharynx and gets funneled into the windpipe, or trachea. And most of the time, the food and water we ingest passes through the pharynx and gets funneled into the esophagus, and then down to the stomach. But not always. Sometimes, food or water wind up in the windpipe, as anyone who has ever inhaled cracker crumbs can tell you.

Sometimes, larger pieces of food get inhaled into the windpipe this way and get stuck there, where they block breathing. In these cases, if the Heimlich maneuver or some other means of removing the blockage isn't performed very quickly, then the victim will die of asphyxiation. There are hundreds of cases like this each year. Many of these result in the sudden premature deaths of otherwise-healthy people.

This shows what happens when things go just a little bit wrong. Food winds up in the windpipe, where it blocks the flow of air. This is fatal unless remedied quickly.

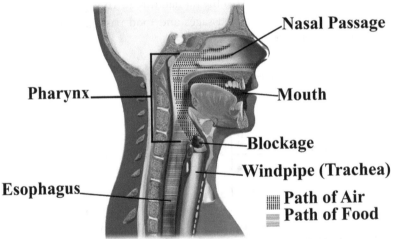

Figure 15.2   Tracheal blockage due to inhalation of food.
This results in an inability to breathe.

A better-designed system would keep the tubes for air and food separate, to avoid unnecessary fatalities. If we were designed, why did the Designer do this job so badly?

Or is it that the Creator likes other animals better? There are creatures in which the air passages and food passages are entirely separate.

Figure 15.3   A whale spouting.

This is a picture of a whale spouting. The mist you see here is caused by the exhalation of air from the whale's blowhole. The blowhole is really its nostrils. The whale's respiratory system is completely separate from its digestive system. This means that a whale, unlike a human, can't choke on its food by inhaling it.

If the Creator could do that for whales, I don't know why he couldn't do it for us.

Chapter 16

The Handy-Dandy Intelligent Design Refuter, Part 2

Just when ID thought it was safe to go back in the water, here's part two of the Handy-Dandy Intelligent Design Refuter.

"WHAT IF WE JUST DON'T RECOGNIZE GOOD DESIGN WHEN WE SEE IT?"

When confronted with evidence of bad design, ID promoters usually hedge, and simply say that we human beings don't necessarily understand what the Designer is trying to do, so what appears to be bad design to us may actually be good design, but we aren't knowledgeable enough to understand it. This boils down to the old philosophy that it isn't given to us to know what God's plan is, or put more simply, "God works in mysterious ways."

Unfortunately for ID promoters, as soon as they bring up this idea, they lose. Why? Because as soon as they say that, they're talking about religion. Even if they substitute the words "the Designer" for "God," it is still admitting that by definition, you can't make testable predictions based on this idea. If you can't make testable predictions based on it, then as science, it's nonexistent. As a religious argument, it may satisfy some people, but it's not science. Once again, they are insisting that acceptance of ID requires *a leap of faith*.

"THIS IS ABOUT ACADEMIC FREEDOM"

The phrase "academic freedom" means the freedom of *college professors* to teach subject material in the way that they think best, even if it is controversial. However, the ID lobby pretends that teachers in American public elementary schools, middle schools, and high schools have academic freedom, which they don't. Academic freedom is something that public school teachers don't have, and have never had. They are required to teach what the school board tells them to teach, using the textbooks that the school board picks out for them. This is why people spend so much time arguing over which textbooks to use, and about who sits on the school board, and what decisions they make. So in arguing that public school teachers must be able to teach ID in order to protect their academic freedom, ID proponents are, as usual, arguing about something that doesn't exist.

Even college professors do not have the freedom to teach anything they want any way that they want. When I am teaching human anatomy and physiology, I am expected to cover specific topics, and I am expected to give scientifically valid information. I cannot simply decide to teach anything I want. For instance, I cannot simply decide to teach the works of Mark Twain instead of human anatomy and physiology in my anatomy and physiology course. If I did, I could lose my job. Not only is Mark Twain not on the anatomy and physiology course description, much of Twain's work is fiction, and I teach science. In the same way, college professors are not free to teach ID in a science course, because it is fiction.

I can also be fired for stating things that are verifiably untrue. So although I may be allowed to state controversial opinions about facts, I have to stick to known facts. So, for instance, I cannot expect to keep my job if I state in class that the sun revolves around the earth, because that is contrary to known facts. ID is contrary to known facts.

"Academic freedom" also does not mean freedom from the need to provide solid evidence in science. It also does not mean freedom from criticism, and does not mean that scientists aren't allowed to reject ID because it is both silly and is lacking in solid evidence.

However, the ID lobby tries to change the subject to academic freedom whenever they can, hoping to cover up for the fact that there is no experimental evidence in favor of ID.

"THIS IS ABOUT 'CRITICAL THINKING'"

Lots of anti-evolution laws masquerade as big-hearted attempts to teach school children "critical thinking" skills.

However, teaching kids to "weigh the evidence" puts kids in the position of being *expected* to believe that there is legitimate evidence on both sides, even when there's not. It also tends to frame the argument as the issue having *exactly* two sides, even though there are as many creation myths as there are world religions, and there are over 4,200 of those.[1]

Since there is no scientific evidence for ID, pretending that there is is nothing more than a bald attempt on the part of the ID lobby to gain legitimacy.

In addition, teaching kids to weigh evidence and decide for themselves is rarely done in elementary or middle school, and only sometimes in high school. Yet these anti-evolution laws want to start kids "weighing the evidence" as early as possible.

Isn't it interesting how these "critical thinking" statutes generally only apply to evolution, and occasionally to global warming, but rarely to anything else? I haven't heard of any "critical thinking" laws being proposed to require the examination, of, say, quantum theory in physics. Or whether or not the planets actually go around the sun. Or whether or not cutting taxes actually boosts the economy. Or whether or not owning guns actually makes people safer.

There is far better evidence for evolution by natural selection than there is for supply-side economics, yet schoolchildren are expected only to concern themselves with the "evidence" against evolution. In other words, the people pushing these statutes don't really care about critical thinking at all. They only care about getting their religious ideas pushed into the classroom any way that they can.

"IT DOESN'T HAVE TO BE PERFECT TO BE DESIGNED"

Some ID promoters argue that something doesn't have to be perfect to be designed. However, something doesn't have to be perfect to be evolved, either, and all the scientific evidence is in favor of evolution.

1. Kenneth Shouler, *The Everything World's Religions Book: Discover the Beliefs, Traditions, and Cultures of Ancient and Modern Religions,* 2nd ed. (Avon MA: Adams Media, 2010), 3.

What's more, as soon as ID promoters make that argument, they have gone out of the realm of anything that's testable.

I say that anybody powerful enough to create all the biological organisms on earth certainly ought to have done better than this Designer seems to have done. We're not talking rocket science, here. We're talking about things that are really stupid and easily avoidable. For instance, having the two tubes of the throat intersect is criminally bad design, and completely preventable.

The fact that I show other species that got better body parts kills ID all by itself. It means that the Designer would have to be unaware of his own work, yet somehow aware enough to have created all the species on earth. Evolution, by contrast, is an unconscious process. This naturally leads to a wide variety of body parts that may be good, bad, or indifferent.

What's more, there has been plenty of time for improvement, but obvious fixes have not been done. Even the Young Earth creationists will allow that the earth has been around for at least six thousand years, yet these bad designs have been allowed to persist, rather than being re-engineered in some simple and obvious ways. Even Microsoft could iron out some flaws in six thousand years.

And finally, perfection is the only way to *prove* design. ID promoters can talk about the design inference all they like, but the rank imperfections in human and other bodies is evidence to the contrary.

INTELLIGENT DESIGN'S SELF-REFUTATION

Intelligent design also has a way of refuting itself. Since it really doesn't have any idea how the suggested Creator actually works, it has problems with basic coherence. Here are a few of its many problems.

Intelligent Design's Problem with Extinction

More than 90 percent of all the species that have ever lived on earth are extinct. Why did the Designer make so many forms during the Cambrian period, for instance, only to have them go extinct? Why were they replaced with later forms? Why have mass extinctions taken place repeatedly throughout history? The most famous one is the Cretaceous-Tertiary extinction, which is the one that wiped out the dinosaurs, as well as many other plants and animals at that time, but it is only one of many. If this

Designer is truly intelligent, it is not clear why he would waste all this time and effort. It is also not clear why he would have so much trouble making things work.

Or is the Designer like a kid with a sand castle, who just builds organisms up in order to kick them down and start over?

Intelligent Design's Problem with New Organisms

Throughout history, both following periods of mass extinction, but at other times as well, new types of organisms have come into being. Does this mean that the Designer shows up on a regular basis, waves his magic wand, and plunks some new organisms down among us? Do ID promoters know his timetable? When do they expect to see him next?

For instance, plants with flowers. They didn't always exist. In fact, they first started showing up during the Cretaceous period, about 130 million years ago. Before that time, you had ferns, and other plants without flowers. Then flowers came into existence, and many new species developed. Why were flowers invented then, and not sooner? Were flowers added later by the Designer, sort of like cruise control on a car?

What's more, it's not all in the ancient past. Every new infectious disease is caused by a new disease-causing organism. The Human Immunodeficiency Virus (HIV) is a good example of a new organism. It causes AIDS.

Intelligent Design's Problem with Transitional Species

Throughout history, there have been organisms that were midway between an earlier species, that had one set of characteristics, and later species that had different ones. Classic examples include *Archeopterix*, a reptile with teeth, but that had wings like a bird's wings and feathers, and *Tiktaalik*, a walking fish that lived about 375 million years ago, and is an intermediate form between fish and amphibians. It even has wrist bones and simple fingers, which are not usual for fish.

How do ID promoters know when something is a new species? What distinguishes it from old species that look a lot like it and appear to be its ancestors? How can they tell newly created species from slight variations in old ones? When is a new feature considered to be a new feature, and when is it just a modification of an earlier feature?

Intelligent Design's Problem with Being a Theory

A real theory of ID would address and predict how often the Designer creates new species. It would address both mass extinctions and the extinctions of individual species. It would address what a new species is, and how it can be distinguished from species that appear to be its ancestors. It would address what we should expect the Creator to do next. ID promoters have done none of these things. They simply claim that something other than evolution took place, but they won't say what it is.

Intelligent Design's Problem with Product Recalls

If the Designer shows up on a regular basis, why doesn't he do product recalls? We could have been given internal testicles, since they've been made compatible with warm body temperature in birds. We could have been given a better birthing system, since the fatality numbers for the current one are so irresponsibly high. We now know that there are better ways of designing all these systems, so why haven't our old, clearly problematic features been redesigned? Why haven't we gotten something better? Even Detroit does recalls. Why can't God?

INTELLIGENT DESIGN'S ANSWERS ARE RELIGIOUS ANSWERS

What predictions for the future would ID theory make? When should we next expect to see the Creator, and what would be on his to-do list? And why are our testicles, eyes, and other features so poorly designed, even when there are easy ways to make them better?

The answer most often heard from ID promoters is simple. They say that we cannot know what the Creator has in mind, so no predictions can be made. They also say that we cannot know whether or not our seemingly poorly designed features are somehow better, in some way that we are not equipped to understand. Unfortunately for their claims to being science, this is a religious argument. It boils down to the old saw that "God moves in mysterious ways."

As a religious argument, this may satisfy some people, at some times. But as a scientific argument, it's nonexistent. There is no way you could ever make a prediction based on that argument, so it's not science, plain and simple. As soon as you hear an ID promoter making this argument, remember—they just lost.

Chapter 17

Irreducible Complexity, the Design Inference, and Geological Formations

Two of ID's favorite arguments are Irreducible complexity and the design inference. I'll explain each one in turn.

IRREDUCIBLE COMPLEXITY

Irreducible complexity is an argument that ID promoters love. The guys who really push this are Michael Behe and William Dembski.

They say that some features of biological organisms are so complex that if you remove a single part of these features, then the whole system breaks down and it doesn't work anymore. They then say that the fact that removing a single piece from these complicated systems makes them break down somehow proves that the organisms that have these structures cannot have evolved from simpler organisms.

I say that they can't tell the difference between simplicity and injury.

As I've said before, their logic is like saying that if you have a dog, and you chop the dog's head off, if the dog dies, then this proves that it couldn't have evolved from a simpler organism.

The logic is lousy, but that doesn't stop ID proponents from using it.

There are many cool things that ID promoters have claimed over the years are irreducibly complex. As each one has been refuted, they've moved on to another one, at least for a while. Then they've gone back to claiming an earlier one, because they still think that they can fool people with it.

So here are the things that ID says are so complex that they couldn't have evolved from anything simpler: the human blood clotting pathway in particular and our biochemical pathways in general, the cellular flagellum, and the human eye.

Unfortunately, rather than being examples of irreducible complexity, all these things are examples of how ID promoters don't do their homework. They are all reducible and often obviously badly designed as well.

For example—if our blood clotting pathways are so well designed, why do we die of blood clotting disorders like hemophilia? If our biochemical pathways are so well designed, why can't we manufacture vitamin C like other animals do? We can die of scurvy as a result of this poorly designed biochemical pathway. These fatal problems could have been prevented.

The cellular flagellum can be reduced, yet ID promoters call it irreducible. Our eyes are both badly designed and thoroughly reducible. This has been known for many decades.

In upcoming chapters, I'll talk about blood clotting, the flagellum, along with its shorter version the cilium, as well as eyes.

THE DESIGN INFERENCE

Another phrase that ID promoters often associate with irreducible complexity is design inference. The design inference means that if something looks like it might have been designed, then they infer that it must have been designed. Another way of putting it is to say that if something looks like it might have been designed, then they assume that it *must* have been. The design inference might better be called the design assumption.

Intelligent Design

So if a biological organism is complex, then ID promoters tend to see both irreducible complexity and the design inference. When ID promoters look around the world, they look for those two things. If something looks complex, and it's not immediately obvious how it could have occurred without divine intervention, then they infer that there must have been divine intervention, in other words, design. Also, if a structure or system breaks when a piece of it is removed, then it must have been designed as well.

What ID's promoters refuse to admit is that some things simply look as though they were consciously designed, when they are really just simple, naturally occurring phenomena. And if they break when mangled, that doesn't mean that God made them.

Take this arch, for instance.

Figure 17.1   This natural arch is in the Sahara Desert.
It is over six feet high, and its thinnest leg is only six inches thick. If you remove even a small part of the thinnest leg, it will collapse, so it is irreducibly complex. It was formed by wind erosion.

NATURAL ARCHES This arch was made entirely by natural forces. It inspires wonder and awe, and primitive people often assumed that such awe-inspiring structures must have been made by divine forces. ID promoters would see both design inference in this structure, since it is not obvious how it was formed, and irreducible complexity, since it will fall over if you remove one leg. But it was not designed by anybody. It came into being by erosion and compression. Two physical, uncaring, unintelligent forces.

Here's another one.

Figure 17.2   Another natural arch.
Look, a bridge! This must have been designed, right? Wrong. This bridge spans the Ardèche
river in France. Although it is nearly 200 feet wide and nearly 150 feet high, and has no
visible support from below, it is a natural formation.

What could be more beautiful than this natural arch? It is irreducibly complex, since you cannot remove certain parts of it without its being destroyed. In fact, if you remove the middle, it would collapse just the way human-made arches do. What's more, it looks as though it was designed for the pleasure of human boaters, or as a bridge to help travelers cross the river. It even looks like a Roman arch, that is, the kind of arch that the ancient Romans used to build their bridges and aqueducts—structures that have lasted for thousands of years and that were definitely made by human beings. So this would appear to be a clear case for the design inference. It is not obvious how an arch like this could have come into being through natural forces. But it did.

ROMAN ARCHES, NATURAL FORCES, AND ART IMITATING NATURE In fact, the reason it looks like a Roman arch is because the natural forces that made this arch are the same natural forces that the ancient Romans learned to exploit when they made their bridges.

This remarkable natural bridge came into being because stone under compression resists erosion more than stone that's not being compressed. In the case of the Roman arch, the ancient Romans figured out how to balance stones so that the compression between them would hold up the arch, and allow air space underneath. In the case of this natural arch, nature did the same thing by eroding away the less compressed parts of the stone, leaving a formation that looks as though it was intelligently designed by the Romans.

In the case of the arch in the Sahara, the compressed parts of the stone resist wind erosion more than the less compressed parts. So as the legs get thinner and thinner, the stone in them gets more and more compressed, which helps defend them against further erosion.

But both natural arches, though marvelous structures, were not designed.

TORPEDOES IMITATE NATURE, TOO The natural arch is a good example of why many biological and human-made structures look similar—the same physical forces are at work on both.

This is also why penguins, dolphins, sharks, and torpedoes all have similar shapes. They are all subject to the same natural properties of water. Anything that travels quickly through water will tend to have that same shape—in the case of torpedoes because humans have tested shapes to find out what shape was most effective. In the case of penguins, dolphins, and sharks, the ones who were more nearly that shape could travel more quickly with less energy, and so tended to survive and reproduce. It does not mean that penguins, dolphins, and sharks were designed simply because torpedoes were. Thinking otherwise is a failure of the imagination.

Figure 17.3   Streamlined shapes.
From top to bottom: penguin, torpedo, shark, whale.

Here are some other wonders of nature that were not designed, even if they look like they have to be. They have many features of design such as being unique, irreducible, and sometimes symmetrical, smooth-sided, or even tunneled through solid rock. Despite these compelling and often beautiful features, they were formed by unconscious natural forces. They were not designed.

Figure 17.4 A unique formation.
This rock formation is utterly unique. It was not designed.

UNIQUE This complex geological formation is different from all others. Its distinctiveness does not mean that it was designed, even though uniqueness is often used as a sign for the design inference.

If we knock out the bottom of this formation, it will fall over, so by ID's definition it is also irreducibly complex.

Yet we know that it was made by natural, not supernatural, forces. The same is true of biological systems, like the blood clotting system, the flagellum, and the human eye. They may be unique and distinctive but that doesn't mean that they were designed. They may collapse if pieces of them are removed. That means that they are irreducible in their current state, but it doesn't mean that they were designed.

Figure 17.5   A symmetrical formation.
Symmetry and a seemingly human form do not make this formation in Utah the work of a Designer.

Symmetrical This formation is symmetrical, and even looks like it could be an artistic rendering of a human being.

These formations emerge when there's a top layer of rock that includes some hard and heavy rock combined with more easily eroded rock underneath. The hard rock on top compresses the soft rock underneath it, making it less easily eroded than the uncompressed rock around it. The uncompressed rock erodes away in the wind.

The symmetry comes into being when the rock erodes away on one side, which makes the remaining rock on that side become more compressed under the weight of the heavy rock on top. Then, since the rock on the other side is less compressed, it erodes more, until enough rock is removed from that side that the remaining rock is as compressed as the remaining rock on the other side. This leads to a roughly symmetrical shape, with the heavy rock very noticeable on the top, making it look like a human with a head of hair. In fact the French name for these formations is *Desmoiselles Coiffées*, or girls with hairdos. In English, we call them hoodoos.

A ROUND-SIDED TUNNEL IN SOLID STONE This is a long, marvelous tunnel with curved, fairly smooth walls that runs for 600 feet through *solid rock*. You can walk through it. It appears to have been carved out of this unyielding rock by conscious designers with supernatural powers or at least a lot of very heavy equipment and a lot of dynamite. It's not something that primitive human beings could have done. It looks better than a lot of subway tunnels. It is hard to imagine this happening through anything other than conscious design and deliberate work by a very skilled and powerful and intelligent being. Yet it is a naturally occurring phenomenon.

Figure 17.6   A round-sided tunnel in solid stone.
Look! A smooth-sided tunnel through solid rock. This formation is found at Volcano National Park in Hawaii. It was not designed.

It's a lava tube. Lava tubes are formed when there is an eruption of lava that flows down a mountainside. Sometimes the top of this stream of lava will harden, since it is exposed to cool air. The lava inside remains hot and liquid, and continues to flow. Eventually, it drains out of the shell of hardened lava that has formed around it, leaving a hollow tube surrounded by hard, unyielding rock.

The processes that made it were unconscious and uncaring. That doesn't make it any less wonderful. And that is the saddest flaw in the reasoning of the ID movement.

ID's promoters simply can't stomach the idea of anything cool happening without them (or someone they know) being in charge of it.

Do you really want to trust the word of such egomaniacs?

Chapter 18

Irreducible Complexity and Blood Clotting

ID promoters love to point to the human blood clotting system as an example of irreducible complexity. One of the problems that the ID folks have is that they don't read much zoology. Blood clotting is simply a system for stopping blood loss when a blood vessel is damaged. Lots of animals have lots of systems for dealing with this. Simple animals have simple—that is, reduced—systems for doing this. They work well enough. More complicated animals have fancier systems. But the whole mechanism is decidedly reducible.

Here's how it works.

Really simple organisms like sea cucumbers only have to tighten the muscles around the damaged area. This stops blood flow. Our bodies do this, too.

Figure 18.1   Pineapple sea cucumber.
In sea cucumbers, the animal constricts muscles near the injured area in order to reduce blood loss.

In slightly more complex organisms, a plug, or clot, is also formed. There are many different levels of complexity for these clots. For instance, in sand dollars and many of their relatives, the blood cells simply clump together on a temporary basis to form the plug, or clot.

Figure 18.2   Sand dollars.
In sand dollars, a temporary plug is formed when blood cells clump together.

In sea stars, on the other hand, the cells, once clumped, then fuse together and lose their identities as individual blood cells. This is a further level of complexity.

Figure 18.3   Sea stars.
In sea stars, the clumped blood cells fuse together to form a clot, and are no longer individual cells.

In horseshoe crabs, the process goes still further. The clumped-up, fused blood cells that start the clot then make protein fibers that enmesh other cells and make the matrix both bigger and stronger. In fact, the horseshoe crab's clotting system is so powerful that bacteria coming in through the injured area can be trapped in the almost-instantaneous clotting, and rendered harmless in this way.

So horseshoe crabs have a clotting system so effective that it not only clots the blood and stops blood loss, it clots out infectious invading bacteria at the same time. The whole process takes very little time. A multimillion-dollar industry has been developed based on using horseshoe crab clotting proteins to test for bacterial contamination in injectable and intravenous drugs used by humans.

Figure 18.4　Horseshoe crab.
In horseshoe crabs, the blood cells form a clot and also spread out fibers. These fibers increase the size and scope of the clot and allow it to trap invading cells.

So horseshoe crabs got a clotting system so effective that infections from wounds are stopped almost immediately. In the days before modern medicine, humans died regularly from infected wounds, often slowly and painfully from gangrene. *Why did the Designer give the horseshoe crabs' life-saving blood clotting system to them, but not to us?*

Meanwhile, animals that have backbones, like us, also have systems with blood that forms both clots and fibers. These functions are similar to those found in invertebrates. But the chemicals that we use to form the fibers are completely different from the chemicals that horseshoe crabs use; the clots form much more slowly, and invading bacteria are not enmeshed in an almost-instantaneous clot.

Irreducible Complexity and Blood Clotting

Figure 18.5   Vertebrates.
Vertebrates like zebras and humans have blood that both forms clots and makes fibers.
But the chemicals that make the fibers in our blood clots are completely different from the
chemicals that do this in horseshoe crabs.

But in all these different animals, the body loses less blood than it would have if the mechanisms for stopping blood loss didn't exist.

So all these different levels of complexity in systems to stop blood loss are at work, today, in the animal kingdom. Some systems are as complex as ours, and some are much simpler. They all work. That is, they help the animal to reduce blood loss caused by injury. What's more, these animals use different chemicals in their blood to produce the same effects that we have in ours. In other words, not only are these systems reducible, they can and do exist in many different forms—that is, they evolved many different times.

The human blood clotting path is individual in that it has some distinctive qualities. This is not surprising, and does not prove or even indicate design. Remember, distinctive individual structures occur all the time in nature, and are formed by well-known natural forces. Horseshoe crabs' blood clotting system has some distinctive features, too.

So it has been demonstrated that blood clotting systems are entirely reducible, since reduced blood clotting mechanisms are found throughout the animal kingdom in many different living animals. This information has been freely available for many decades. It can be found in undergraduate textbooks.

Unfortunately it appears that ID promoters have deliberately overlooked all the information that is available on this subject. Either that, or else they don't know how to read.

What's more, the human blood clotting system is not as fantastic as the ID promoters would have you believe. For more on the *bad design* of our blood clotting system, see Chapter 19.

Chapter 19

Bad Design—the Human Blood Clotting System: It Led to the Communist Russian Revolution

Another thing that the ID folks have deliberately overlooked is the prominent role that our *badly designed* blood clotting system has had on human history. If the human blood clotting mechanism were as marvelous as the ID folks claim it is, then the Russian Revolution probably would never have succeeded.

Think about it. The last Russian czar had an only son named Alexei, who would have become the czar after his father died. However, Alexei had hemophilia—*a blood clotting disorder.* This type of hemophilia leads to frequent uncontrollable bleeding even if the body has suffered no injury at all, not even a small cut. These days, due to scientific advances, we can treat hemophilia successfully. But in those days, no cure or reliable treatment existed.

Alexei's blood clotting disorder led his mother to become overly dependent on a self-proclaimed faith healing monk named Rasputin. She believed that he could successfully treat her son's incurable clotting disorder.

Figure 19.1   Alexei.

Unfortunately, Rasputin was a very disreputable character. He made lots of enemies for himself, and for the poor, deluded czar and his wife. He convinced the czar to personally command his troops during World War I. This had terrible consequences for the army, and for the Russian war effort. What's more, the czar's wife, who was Alexei's mother, then became the acting ruler of Russia, making all the decisions about how to conduct the war and run the country. With the czar away from home, Rasputin gained even more influence over her. Soon, the country was in chaos. Eventually, the royal family's inept handling of both the country and the war lead to the communist takeover of the Russian government, the formation of the Soviet Union, and eventually to the subjugation of Lithuania, Latvia, Poland, Hungary, Czechoslovakia, Bulgaria, and other Eastern European countries by the Soviet Union.

Figure 19.2 Rasputin.

If Alexei hadn't had his blood clotting disorder, then the Russian royal family would not have become dependent on Rasputin, and wouldn't have taken his bad advice.

Without Rasputin, they probably would have better handled the problems of their country and the First World War.

So, the *faulty human blood clotting system* was a leading cause of the Communist takeover of Russia and much of Eastern Europe. Does the Creator really want to take responsibility for the seventy-two years of

Communist totalitarianism in Russia, and the subjugation of most of Eastern and Central Europe?

Is the Creator a Communist? *Are Intelligent Design promoters Communist sympathizers?*

The Hemophilia Is Sex-Linked

What's more, this form of hemophilia is inherited genetically, and only men can get it. This is called being sex-linked. So—*does this Creator prefer women to men?*

Chapter 20

Irreducible Complexity, Flagella, and Cilia

FLAGELLA AND CILIA

ID promoters love to talk about the little hairy extensions that many cells have, which are called cilia and flagella. Cells that have nucleuses have both long threadlike extensions, called flagella, and shorter ones called cilia. These cilia and flagella have the same internal structure, but cilia are much shorter.

Bacteria are a very different kind of organism from us. They are one-celled organisms, and the cells are much smaller and do not have nucleuses. Bacteria also have threadlike extensions called flagella, but these have a completely different structure from the flagella found in cells that have nucleuses.

The first thing that ID folks do wrong is they don't distinguish a cilium from a flagellum, and the second is that they don't know the difference between bacteria and everything else.

Thing One: Bacteria Don't Have Cilia

So the first thing you need to do when talking to ID promoters on this subject is to ask them whether they are talking about the flagella found on bacteria, or the flagella found on cells with nucleuses, such as the human sperm cell. These kinds of cells are completely different, and so are their flagella.

Humans and bacteria are not closely related, and their flagella are entirely mechanically different. Saying that they are the same would be like saying that the human arm and a jellyfish tentacle are the same. So always ask—which flagellum is ID talking about?

Sometimes ID promoters talk about the cilia on bacteria. For the record, cilia *don't exist* on bacteria. So if you ever hear an ID promoter talking about the bacterial cilium, you know he's off base before he even gets started on anything else. Cilia only show up on cells that have nucleuses, which bacteria don't. Cilia are shorter versions of flagella, but with the same structure. Sometimes, cilia will cover large portions of their cell's surface. But they never show up on bacteria at all.

The Cilium — "Irreducible Complexity Squared"?

So, in his book *The Edge of Evolution: The Search for the Limits of Darwinism*, ID promoter Michael Behe tells us that cilia and flagella in cells with nucleuses can't be made without a particular cellular system called IFT, which stands for intraflagellar transport. He says that IFT is universally required for cilium and flagellum construction.

He also claims that both the cilium (or flagellum) itself and the IFT are irreducibly complex. Since both the cilium and IFT are (according to him) irreducibly complex, this would make a completed cilium even more irreducibly complex, since even the system that makes it can't be reduced, and if that system is removed, then the cilium can't be constructed. He christens this setup "irreducible complexity squared"![1] He thinks this is a great argument.

There's just one problem. Cilia and flagella can be made without IFT. So he simply got the science wrong.

It gets worse. In the same book, Dr. Behe also states that the parasite that gives us malaria is a great example of intelligent design. He acts as though he has studied malaria with great love and care. However, the malarial parasite, which has a nucleus and is not a bacterium, does build a flagellum, *but it doesn't have IFT!*

1. Behe, *The Edge of Evolution: The Search for the Limits of Darwinism* (New York: Free Press, 2007), 93.

So the very organism that Dr. Behe claims to have studied with such great care that he wrote a large portion of a book about it, is actually a living refutation of his claims about irreducible complexity. It has a system for building a flagellum that is reducible, which Michael Behe says doesn't exist. What's worse, Dr. Behe doesn't seem to know about this.

In my opinion, Dr. Behe should read about the organisms he writes about, before he writes about them. In doing otherwise, he has failed inexcusably. By claiming that all cilia and flagella require IFT in order to be made when this isn't true, Dr. Behe has failed inexcusably.

... or Inexcusable Failure, Squared?

So, to recap: Dr. Behe says that all flagella and cilia in cells that have nucleuses require IFT. They don't. For Dr. Behe, that's *inexcusable failure*.

Furthermore, malaria itself, the very organism that Dr. Behe claims is proof of ID, doesn't have IFT. Dr. Behe apparently doesn't even know this. This, too, is *inexcusable failure* on his part. That's *inexcusable failure, squared!*

Chapter 21

Irreducible Complexity and the Human Eye

Now I want to get into an example that the ID people use repeatedly, hoping to show that human anatomy is so complicated and so perfect that it couldn't possibly have evolved from anything simpler. The human eye.

ID types say "What good is half an eye?"

Well, I'm going to show you what good half an eye is, as well as a quarter of an eye, an eighth of an eye, a sixteenth of an eye, and a whole lot less. I will also show you that there are creatures alive today that have these eyes and benefit from them.

In fact, it is clear that vision has evolved separately many times in the animal kingdom, and it is clear that some animals have eyes that are actually better put together than ours. Which leads us to the interesting question of "Who does the Creator like better—us or squid?" But I'll get to that later.

First, let's start with the human eye.

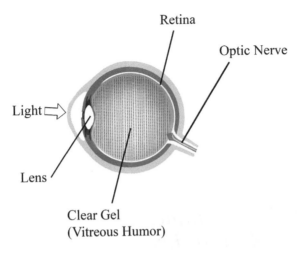

Figure 21.1 The human eye.

Here's a picture of the human eye. The more important features include the enclosed shape (that only allows light in through a tiny opening at the front), the lens, the optic nerve, and the light-sensitive retina at the back. The retina is the place where the cells that respond to light, the photoreceptor cells, are located. The whole point behind having an eye is the photoreceptor cells. Everything else is just accessories.

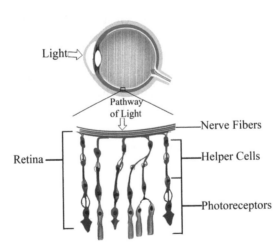

Figure 21.2   Diagram of the human retina.
This diagram shows the pathway that light must take to reach the photoreceptors. Note the nerve fibers and multiple layers of cells through which the light must pass before making contact with the photoreceptors.

Figure 21.2 shows a close-up of the retina. One odd feature of the human retina is this: the nerve fibers taking the information from the retina to the brain go in front of the photoreceptors, and block some of the light. I'll get back to that later. But please notice for now where those fibers are located.

Now, let's look at clams. Here's a picture of an eye. It belongs to the bivalve *Pecten maximus.* All clams and their relatives are bivalves.

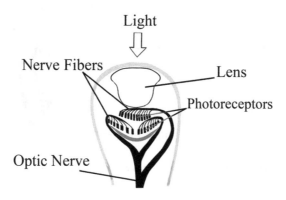

Figure 21.3   The closed-lens eye of *Pecten maximus.*

You can see that it has a lot of the same features as the human eye—the enclosed shape, the lens, and the retina at the back with the wiring running in front of the photoreceptors. It looks a lot like ours.

Here's another eye, belonging to *Cardium muticum*, another bivalve.

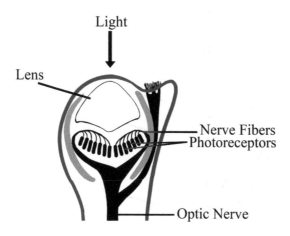

Figure 21.4   The closed-lens eye of *Cardium muticum.*

It, too, has similar features to the human eye.

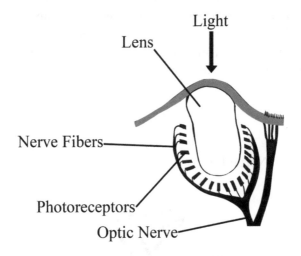

Figure 21.5 The closed-lens eye of *Tridacna maxima.*

But here we have *Tridacna maxima*. Another bivalve. And look! In this case, the wiring is *behind* the photoreceptors. This probably means that vision in this line evolved *separately* from that of the others. Or else it means that the Creator likes this clam better than the others, and better than us.

Now things start getting really interesting. Here's an answer to the question, "What good is half an eye?"

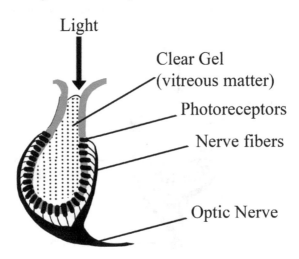

Figure 21.6 The pinhole eye of *Lima squamosa.*

Here is the pinhole eye belonging to *Lima squamosa*. Look! No lens! This is a much simpler structure. But this eye still helps the animal to gather information about the outside world.

It has the wiring behind the photoreceptors, too.

Here's an example of even less of an eye.

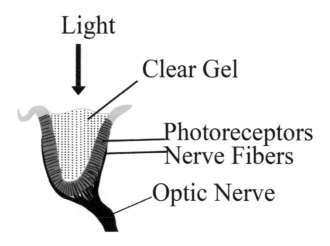

Figure 21.7   The eye pit of *Arca barbata*.

This is nothing more than a pit, with a few photoreceptors at the back.

It belongs to *Arca barbata*, another bivalve. It too has wiring behind the photoreceptors. Despite its obvious simplicity, it is still a visual organ. It "sees." That is, it perceives light and sends that information to the animal's nervous system. It is a lot less than half an eye—and it works. It provides useful information to the animal that aids its survival.

Keep in mind, all these animals are related. *And they are all alive today.* So at present, all these eyes are out there, doing the job that eyes do, even though some of them are half an eye, or a quarter of an eye, or possibly even better eyes than ours. What they have in common is that their possessors are better off with them than without them. That's the mark of evolution—*not* perfection, just *increased survival.*

Now, in case you thought that all that diversity was limited to clams, let me show you another lineup.

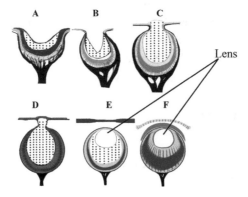

||||-Clear Gel (Vitreous Matter)

Key to gastropod pictures
A) The eye pit of *Patella* sp.
B) The eye cup of *Pleurotomria beyrichi.*
C) the pinhole eye of *Halitosis* sp.
D) the closed eye of *Turbo Creniferus.*
E) the lens eye of *Murex brandaris.*
F) The lens eye of *Nucella lapillus.*

Figure 21.8   Gastropod eyes.

Now you're looking at snail eyes. All these different eyes, with all their degrees of complexity, are owned by gastropods—snails and their relatives. They are all much more related to one another than any of them are related to us. Here, again, you can see eyes ranging from ones that are fully as complex as ours, to eyes that are little more than pits with light receptors in them. Again, they all work.

And now, in case you thought I had shown you the simplest eyes imaginable, I'm now going to show you one that's even simpler.

This is a picture of a *Euglena*.

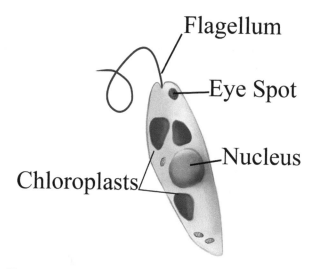

Figure 21.9 Unicellular *Euglena* with eye spot and chloroplasts.
The eyespot, a cluster of pigment, helps the organism sense and respond to light. The organism then moves toward light using its flagellum for locomotion. This way, the chloroplasts are properly positioned for photosynthesis. This allows the cell to make food.

A Euglena is made up of only one cell. Yet it has an eyespot—that is, a spot that senses light. The creature can then move toward the light, and make food from light. So the eye "works." It senses light and helps the animal to survive. Make a note of this—the eye does not even take up an entire cell. It is nothing more complex than a patch of pigment. But it works.

PAY ATTENTION NOW: There is not a single stage in the evolutionary process in which animals have eyes, but the eyes are too simple to work. Biologists have known this for decades. The facts about these animals and their eyes have been published openly for decades. Why haven't the ID people heard of *all these* animals?

I suspect that it's because they spend a great deal of time advertising themselves, and very little time studying zoology, where all this information is freely available.

Perhaps they prefer spending time in front of cameras to spending time in libraries.

Perhaps they know they'll lose if they actually start studying science.

Learning biology is hard. Maybe it's more fun for them just to feel aggrieved.

Whatever their reasons, they seem to be far more interested in talking than they are in getting their facts straight.

What's more, ID proponents insult *our* intelligence. They think we can't tell the difference between "half an eye" and "a simple eye." For clarity, I'll illustrate the difference.

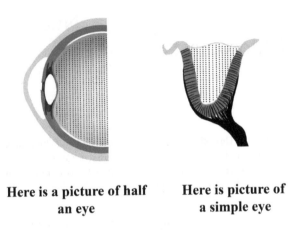

Here is a picture of half an eye　　　**Here is picture of a simple eye**

Figure 21.10　Half an eye and a simple eye.

In the animal world, there are many simple eyes. There are also simple hearts, simple brains, and simple digestive systems, to name just a few examples. They all work.

So now I've addressed the issue of irreducible complexity. I've shown you that eyes can indeed be reduced and still function very nicely, and do, throughout the animal kingdom.

Now let's move on to the whole notion that the human eye is so perfect that it must have been designed, and by an infallible Designer at that. I say that if indeed it was designed, then the Designer would get an "F" in any decent design class.

Chapter 22

Bad Design—the Human Eye

Here's why our eyes are a case of *bad design*.

Try to imagine a camera so poorly designed that the mechanical parts of the camera—the wires and so on—were between the lens and the imaging chip, so they left their shadows on all your pictures. Then imagine having to photoshop those shadows out of every one of your pictures in order to have usable images. Imagine further that all of the wires are routed *through a hole in the imaging chip* to the back of the camera, so that there is a permanent hole in the chip where there can be no image at all. This is the case with our eyes. This is *bad design*.

The part of our eye that is sensitive to light, the part that is like the imaging chip in a camera, is our *retina*.

The first bit of *bad design* in our retina is that there are blood vessels sitting on the surface of the retina, blocking the light. A decent designer could have put those blood vessels behind the retina.

The second bit of *bad design* is that there are nerve fibers that also sit on the surface of the retina, blocking even *more* light. These, too, could have been placed behind the retina, if anyone had been trying to design the eye properly. Unfortunately, no one was.

The third bit of *bad design* is that these blood vessels and nerve fibers all meet in one spot and go *straight through the retina* and into the optic nerve. This is like punching a hole in the imaging chip of a camera. It creates a part of our retina that is so clogged with all this paraphernalia that it cannot see at all. This is known as the *blind spot*, or *optic disc*. If all these blood vessels and nerve fibers had been properly placed behind the retina in the first place, then this hole in our vision would not exist. Putting preventable blind spots into a visual system is *not* the sign of an infallible Creator.

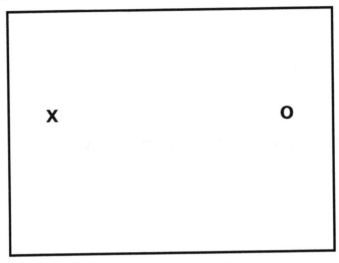

Figure 22.1   Find your blind spot.

Here is my proof, in case you need it. You can find your own blind spot. Just use the figure above, and follow these directions:

Hold the test about eighteen inches from your face. Close your left eye while focusing your right eye on the X. The X should be placed directly in front of your right eye. Move the test slowly toward your face, and keep your right eye trained on the X. When the O located to the right of the X goes away, you have found your blind spot.

So now that you can find your blind spot, let me show you pictures of why it exists. Here is a picture of your retina taken with an ophthalmoscope, showing the blood vessels lying on top of the retina.

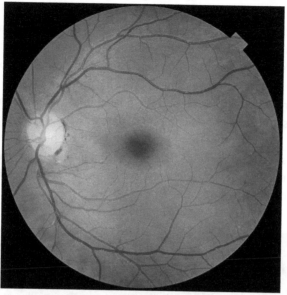

Figure 22.2   Photograph of the human retina, taken with an ophthalmoscope.
Notice the blood vessels lying on top of the retina, which block the path of the light that's traveling to the photoreceptors.

And here is the diagram of your retina, showing the nerves lying on

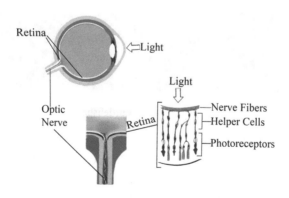

Figure 22.3   Diagram of the human retina and optic nerve.
Notice the blood vessels, nerve fibers, and many layers of cells that are blocking the path of the light that's going to the photoreceptors. The nerve fibers and blood vessels on top of the retina converge before entering the optic nerve, blocking the light entirely. This forms the human eye's blind spot.

top of the retina and blocking the light that should be going to the photo-receptors underneath. This also shows the blood vessels and nerve fibers diving through the retina to get to the optic nerve.

What all this means is that we need more light to see with than we would if our eyes had been designed properly. It also means that your nervous system receives poor quality visual information that then has to be processed several times before the information becomes a good image. If this were a camera, this would mean exposing multiple pictures with all the shadows from the wires on it and the hole going right through it, loading all the lousy pictures onto a computer, removing the shadows of the wires from each one, and combining several until you got one usable image that more or less represents reality. Even then, some information has been lost. This is a cumbersome, imperfect process for a supposedly infallible Creator to have come up with.

One would think that God could do better.

Figure 22.4   Cuttlefish.

And God did.

This is a picture of God's favorite. It's a cuttlefish. It and its relatives, squid and octopi, have eyes with retinas that are built far better than ours. They see better, and in dimmer light than we do. They can also recognize polarized light, a useful ability that we don't have at all.

So—does the Designer like cuttlefish better than humans?

Now take a look at that beautiful eye, so much like ours, but wired so much better, and then look at these drawings.

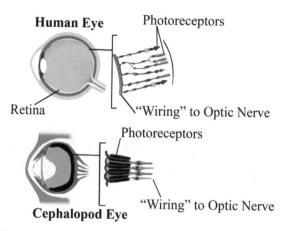

Figure 22.5   Diagrams of human and cephalopod eyes.

These are diagrams of the human eye and the cephalopod eye. Cephalopods do not have backbones and include squid, octopi, and cuttlefish. Notice how the "wiring" that takes the visual information from the photoreceptors to the optic nerve is located behind the photoreceptors in cephalopods, but between the photoreceptors and the light in humans. As I said earlier, this is like having the wiring for a camera placed between the lens and the imaging chip. It makes no sense if our eyes had been designed, but from an evolutionary standpoint it makes sense, because it works well enough and does us more good than harm.

However, I *should* point out that the standards for evolved features can go even *lower* than that. Sometimes they don't even meet the criterion that they do more good than harm. Sometimes the only standard that they meet is that they don't kill us before we reproduce . . . too often.

Chapter 23

Why Denying Science is Bad for You, and Bad for Your Neighbors, Too

WHY DENYING SCIENCE IS BAD FOR YOU

People who deny scientific facts pay a heavy price. So do the people around them. So do people whose governments deny scientific facts. This has caused human tragedy on a major scale.

A good example of this is AIDS in South Africa. South Africa has been very severely affected by AIDS. As of 2006, approximately 18.8 percent of the adult population of South Africa was infected with the HIV virus. The HIV virus causes AIDS. Left untreated, HIV infection leads to death. In 1999, as the epidemic was rising, President Thabo Mbeki announced that HIV was not the cause of AIDS, and refused the donation of anti-AIDS drugs, saying that they were not useful. His minister of health told AIDS sufferers to eat garlic and beetroot instead. This was in spite of international scientific consensus on the subject, and proof that AIDS drugs worked.

The result was predictable. Between the years 2000 to 2005, more than 334,000 South Africans died of AIDS.

The neighboring countries of Namibia and Botswana accepted the free drugs, and their populations did much better.

Among the many people who died in South Africa were babies who were infected with HIV through their mother's placenta. One single dose of an anti-AIDS drug could have saved each of these babies, but they all got

AIDS instead, because Thabo Mbeki believed the folks who deny scientific facts.[1]

AIDS Denialists Are ID Promoters

Interestingly, a number of enthusiastic HIV/AIDS deniers are also enthusiastic promoters of ID. One is Phillip Johnson, a law professor who became a born-again Christian and went on to become the father of ID, and who is a program advisor at the Discovery Institute's Center for Science and Culture. Jonathan Wells, another ID promoter and a senior fellow at the Center for Science and Culture also denies the HIV/AIDS connection. Both Johnson and Wells are enthusiastic public supporters of the leading proponent of denying that HIV causes AIDS, a man named Peter Duesberg.[2] They have written public letters and articles in support of his stance. Thabo Mbeki consulted with Duesberg about the cause of AIDS, and then publicly announced that AIDS is not caused by HIV.[3] As a result, he turned down the free, lifesaving drugs. So hundreds of thousands of adults died, innocent babies died, and many children were made orphans.

This wholesale slaughter of innocents was encouraged and supported by the same irrational guys who now insist that this country believe them about ID. They want to change our school textbooks, too.

Do we want to trust either our health or our children's science educations to these irresponsible men?

Even Bigger Killers Than AIDS, and Evolution Predicts Them

Unfortunately, today, there are even bigger killers than AIDS. For instance, MRSA—that's Methycillin Resistant *Staphylococcus aureus*. This disease alone kills more people in the United States than AIDS does.

1. *Journal of Acquired Immune Deficiency Syndrome,* 49, 4 (December 1, 2008), http://aids.harvard.edu/Lost_Benefits.pdf.

2. See Charles A. Thomas, Jr., Kary B. Mullis, and Phillip Johnson, "What Causes AIDS? It's an Open Question," http://www.duesberg.com/articles/kmreason.html; Eleen Baumann et al., "AIDS Proposal," *AAAS Science,* February 17, 1995, 945–46; P. Duesberg et al., "Rethinking AIDS Letter," http://rethinkingaids.com/quotes/rethinkers.php; and "AIDS 'Denialism' Gathers Strange Bedfellows," the *Vancouver Sun,* http://www.canada.com/vancouversun/columnists/story.html?id=b0cb194b-51d3-4140-88f7-e4099445c554.

3. See Sarah Boseley, "Mbeki AIDS Policy 'Led to 330,000 Deaths,'" http://www.theguardian.com/world/2008/nov/27/south-africa-aids-mbeki.

According to the *Journal of the American Medical Association* (JAMA), more people died in the US from the MRSA infection (18,000 deaths) than from AIDS (16,000) in 2005, and MRSA's numbers are rising faster than AIDS' numbers are in the US.[4]

And how did MRSA arise? It arose from our ignoring evolution. Ordinary *Staphylococcus aureus* (a Staph infection) is a dangerous infection, but one that can be treated successfully with modern antibiotics. Unfortunately, most people either didn't understand evolution by natural selection, or chose to ignore it. So people all over the world abused antibiotics, taking them when they didn't need them, or taking just a few, then leaving the rest in the bottle. Why is the latter a problem? Here's a hint: when you take antibiotics, you need to take the full prescription, because you need to kill *all* the disease-causing bacteria in your system, not just the weak ones. Otherwise, the ones that have a little bit of natural antibiotic resistance will be the ones that remain alive to breed. Repeat this by the millions of generations of bacteria that have reproduced themselves since the dawn of modern antibiotics, and in just the few decades since antibiotics have become widespread, there has been more than enough time for dangerously antibiotic-resistant bacteria to breed up into dangerous populations around the world.

Abuse of antibiotics has become so bad that antibiotics have even been put into cattle, pig, and chicken feed, so that animals can gain a few more pounds before being slaughtered. Meanwhile, our onetime wonder drugs have become less and less powerful because of this abuse. Now, if you go into a hospital with a routine problem, you run the risk of getting an antibiotic-resistant Staph infection (MRSA) while at the hospital, and dying.

Evolutionary theory predicted this problem. ID has predicted nothing. If ID is really a scientific theory, I would like its promoters to tell us what new biological features and problems are going to be created by the Designer. That way we can safeguard ourselves against them.

Deadly Diseases We Used to Be Able to Cure, but Now Can't, Because We Ignored Evolution

In addition to MRSA, there are many other life-threatening microbes that resist antibiotics, including dangerous diseases like tuberculosis, strep

4. On AIDS deaths, see Sarah Boslaugh, ed., *Encyclopedia of Epidemiology* (Thousand Oaks, CA: SAGE, 2008). For MRSA death numbers, see R. Monica Klevens et al., "Invasive Methicillin-Resistant *Staphylococcus aureus* Infections in the United States," *Journal of the American Medical Association,* vol. 298, no. 15 (October 17, 2007), 1763–71.

throat, pneumonia, meningitis, dysentery, typhoid, cholera, and gonor-rhea; and dangerous diseases like malaria and viral hepatitis that resist other modern antimicrobial drugs.

Malaria is the disease that Dr. Behe claims is evidence of "irreducible complexity squared"! Yet, in addition to showing lack of design as discussed in Chapter 20, it also shows clear signs of having evolved its current resistance to modern drugs by way of evolution through natural selection.

All the deadly diseases in my list are things we could cure until recently, but now we can't always. All this is due to the fact that evolution by natural selection was ignored.

Unfortunately, this is a case in which it is not enough to be sensible all by yourself. If other people abuse antibiotics, you can die as a result.

POLIO AMONG THE AMISH AND OTHER PROBLEMS: TRYING TO LIVE WITHOUT SCIENCE FOR RELIGIOUS OR PERSONAL REASONS CAN KILL YOU, AND KILL OTHERS

The Amish, those hard-working, deeply religious people who live in Pennsylvania, Ohio, and elsewhere, do not believe in vaccinating their children against polio (poliomyelitis). Unfortunately, wild polio viruses can live naturally in our environment, and every now and then one can come into contact with an unvaccinated person. That person can get polio. When that person lives in a population of unvaccinated people, an epidemic can get started in that population.

In the early twentieth century, polio crippled thousands of children every year, even in industrialized countries. Then, in the late 1950s, the first effective polio vaccine, inactivated poliovirus vaccine, was introduced. In the early 1960s, another vaccine, oral polio vaccine, was introduced. After that, polio was largely eliminated from industrialized countries.

Except among people who don't believe in vaccination. In 1979, there was a polio outbreak in Amish communities in Iowa, Wisconsin, Missouri, and Pennsylvania. Ten people were left paralyzed by the disease, according to the Centers for Disease Control and Prevention.

In the Netherlands, between 1992 to 1993, fifty-nine non-vaccinated people were paralyzed by polio and two more died. All but one belonged to religious groups that refused vaccination.

In more recent years, vaccination campaigns have wiped out polio in most developed countries. Unfortunately, vaccination efforts in the remaining areas are often hampered by unscientific opposition. For example, in both Nigeria and Pakistan, polio-eradication workers have been attacked after political and religious leaders have said that the vaccinations caused sterility, or are against God's will.[5]

This has lead to upsurges in polio in these regions. If polio is not wiped out there, then it can re-spread to elsewhere in the world. So, living as though scientific research doesn't matter can literally kill you, kill your neighbors, and kill your children.

In developed countries, people sometimes choose to not get vaccines. Unfortunately, in 2003, an unvaccinated man from the US died from diphtheria after travelling to Haiti.[6] This is a vaccine-preventable disease. In June 2015, an unvaccinated child in Spain died of diphtheria.[7] When their son got sick, his parents said that they felt "tricked" by anti-vaccination groups that they had previously admired. The family said that they had not been properly informed by these anti-vaccination groups.[8]

On July 2, 2015, a woman in Washington state died of the measles.[9] She had been vaccinated, but had to take immune system-suppressing drugs due to other health conditions. She was most likely exposed to the measles at a local health facility by an unvaccinated person. The immune-suppressing drugs made it harder for her body to fight the measles infection, so she died. If those around her had been vaccinated, she would not have caught the disease from them, and wouldn't have died.

Again, living as though scientific research doesn't matter can kill you. Even worse, it can also kill other, innocent people.

---

5. Warraich H. J., "Religious opposition to polio vaccination" [letter], *Emerging Infectious Diseases* 15, 6 (June 2009).

6. Centers for Disease Control, "Fatal Respiratory Diphtheria in a U.S. Traveler to Haiti—Pennsylvania, 2003" *Morbidity and Mortality Weekly Report*, January 9, 2004, 52(53), 1285–1286.

7. Jessica Mouzo Quintáns, "Six-year-old boy who contracted diphtheria dies in hospital," *El Pais*, June 29, 2015

8. Jessica Mouzo Quintáns, "Parents of diphtheria-stricken boy feel 'tricked' by anti-vaccination groups," *El Pais*, June 5, 2015.

9. Tara Haelle, "First Confirmed U.S. Measles Death In More Than A Decade," *Forbes*, July 2, 2015.

Chapter 24

Why Reward in Heaven Makes No Difference to Human Behavior, but Good Institutions Do

As I discussed in Chapter 10, most religions try to explain the world, and they set out a moral code. Explaining the world gives the people who follow that religion some sense that they can control their world.

So, for instance, the world may be explained as being full of gods that have special powers over rain, wind, sickness, and so forth. People can get a feeling of control over the world by doing the appropriate religious rites. For instance, someone with a sick child can make an offering to the god in charge of illness, in the hope that the child will be spared. Me, I prefer modern medicine.

Often, a religion's authority for setting out its moral code is based on the idea that it has explained the world. For this reason, many religions resist scientific explanations of the world, because they feel that this undermines their authority.

They also feel, oddly, that having a scientific explanation of the world instead of a creation myth somehow undermines having a decent moral code of any kind. If religions overcame the idea that they cannot speak about morals unless their creation myth is accepted, and simply spoke directly about morals and values, they would speak with much greater clarity and dignity.

Even more oddly, many religions believe that if you have a moral code, then you are *required to believe* their creation myth, as if the two have much of anything to do with one another. They believe that you have to accept their version of how the world got started in order to have decent morals. They believe this though it makes no sense.

In fact, many people in the religious Right say that without religion, people will have no morality. Specifically, many claim that people will not behave well toward other people unless they are threatened with hell as a punishment for bad behavior toward others, and promised a reward in heaven for good behavior toward others.

This ignores many things. First, it ignores a depressing level of bad behavior by many religious people throughout history. Although it is sometimes claimed that religion makes people behave better, this assertion is not based on evidence.

Second, it ignores the awful behavior that has sometimes been encouraged by religion such as torturing and killing heretics, and wars based on requiring that a specific religion be accepted by the entire population of a region. Examples of the latter include the Crusades and the Thirty Years' War.

Third, it ignores the fact that Christian denominations often have loopholes for bad behavior. For instance, Catholics believe that they can sin and still expect to get into heaven, because they can go to confession, confess everything, and then say the prayers required of them. After that, Catholic theology says that they can get into heaven just as though they had never committed those sins. They may have to go through purgatory first, but they'll get there in the end. Insincere lip service of true repentance has often been tolerated by the Catholic Church. For instance, it was only in 2014 that the Pope excommunicated members of the Mafia, an organized crime ring known for drug running, murder, and extortion.[1]

Many conservative Protestant denominations expend a lot of effort trying to identify and avoid sin. But many of the actions that they call sins are actions that have no human victims. For instance, adults engaging in consensual sex may have no human victims at all. Likewise, blasphemy definitely has no human victims. Yet these sorts of "sins" are often the ones that are most preached against by conservative Christians. Further, many denominations say that people will go to heaven if they repent of their sins

---

1. Stille, Alexandra, "The Pope Excommunicates the Mafia, Finally", *The New Yorker*, *Daily Comment*, July 24, 2014, http://www.newyorker.com/news/daily-comment/the-pope-excommunicates-the-mafia-finally.

and accept and trust Jesus as their Lord and Savior. None of these ideas leads to the conclusion that good behavior toward other people is necessary for getting into heaven, or for avoiding hell.

In other cases, if a church says that it believes in salvation by grace alone (*sola gratia*), then this means that a person's behavior during his or her life does not influence God's decision on who gets into heaven and who gets sent to hell. At all. That's what salvation by grace alone means, and it's a central tenet in many conservative Christian denominations.

On the other hand, many liberal Christian denominations reject the idea that hell is a real place. Others believe that heaven, if it exists, is where everybody gets to go after death. Again, these ideas do not lead to the conclusion that good behavior toward others is necessary for getting into heaven, or for avoiding hell.

So, Christians may say good things about other people and do good things for other people. They may say that they are doing it in the name of their faith. But no matter what type of Christians they are, their faith's dogma may not insist that they have to behave well toward other people in order to get into heaven.

On the other hand, "crimes" against dogma may be punishable by hell, even though they have no human victims. Crimes against dogma are things like heresy, apostasy, or blasphemy. This means that good behavior toward other people is less important than what you say or do regarding god. So it is often the case that a faith's theology specifically does not decree that good moral behavior towards other people has much to do at all with a person's going to heaven.

What's Hell Got to Do with It?

Finally, it also ignores the fact that many of the upright people who settled this country did not believe that heaven would be their reward for a virtuous life, nor hell their punishment for a bad one.

In fact, they believed that heaven and hell had *nothing to do with a person's behavior.* Nothing.

What's John Calvin (and the Puritans, and the Presbyterians) Got to Do with It?

You heard me right.

A significant portion of the United States was settled by Puritans and Presbyterians, both of which are Calvinist religions. This means that they believed that heaven was not a reward for good behavior on earth, nor hell a punishment for bad behavior. Why? Because in Calvinist religions, whether you go to heaven or to hell is decided long before you are born, and nothing you do during your life can change it.

That's a Calvinist doctrine.

So, many of the people who settled this country did not believe that they would be rewarded for good behavior in heaven, nor punished for bad behavior in hell.

In the case of Puritans, they believed that only members of the "elect" would go to heaven, and this was decided before a person was born. However, they thought that in life there were signs of proof that someone had been elected. These signs were things like doing good works, having a successful business, having a healthy family, and being an upstanding citizen. So Puritans often tried to behave well, but not in order to get into heaven.

In the case of Presbyterians, they too believe that only the elect get into heaven and that this is decided before you are born, and that nothing you do on earth can cause you to deserve to get into heaven. However, they recommend that you be good and generous to other people for different reasons—to express gratitude for salvation, to follow advice in the Bible about feeding the hungry and clothing the needy, or to behave toward the unfortunate as Christ did.

None of this involves earning points toward getting a good afterlife!

Yet the Puritans and Presbyterians kept to their rather strict moral codes about as well as anybody sticks to their moral code. They built thriving cities, towns, and farms. They put a great deal of effort into making good cities, good institutions, and good societies. They thought that good laws that stopped people from taking unfair economic advantage of others were important, as were providing good schools and taking care of the destitute.

But they did not do this in order to give themselves a good afterlife. They did this all without the benefit of thinking that they were working toward a reward in heaven, or trying to avoid punishment in hell.

We Are Told that Those Were the Good Old Days

The times when they did this are considered by many conservative groups to be the "good old days" in America. Yet, when people were good, they were often good without any consideration at all of how this would affect their chances of getting into heaven.

So it has been proven by American history itself that society can do very well without using threats of heaven and hell to keep us in line.

The Wedge Document is Anti-Christian and Anti-American

What's more, Calvinist Christians believed, and still believe, that humans cannot be perfected, but they can and should be encouraged to behave better by setting up good secular institutions. They believe that better behavior here on earth is possible, and desirable in its own right. Good secular institutions include things like public schools, public libraries, and other public, government-run establishments.

The folks who push ID run directly counter to this Christian doctrine. Here's a quote from the Wedge document:

> Finally, materialism spawned a virulent strain of utopianism. Thinking they could engineer the perfect society through the application of scientific knowledge, materialist reformers advocated coercive government programs that falsely promised to create heaven on earth.[2]

So the folks at the Discovery Institute directly oppose good public institutions that help people to behave better and lead better lives, because they say that this is "utopian." Never mind that many of the Calvinist Christian reformers who set up many of these institutions were the exact opposite of utopians, and never believed that human society could be perfect. They merely believed that it could be better, and that a better society, created through better public institutions, was worth working for.

Evidently, the Discovery Institute wants to take apart all the fine institutions and public works that good people throughout American history have worked so hard to make happen, because making people's lives better is "utopian." This means that the Discovery Institute is anti-American. Now let's move on to scurvy, another example of how we are badly designed.

2. See http://ncse.com/creationism/general/wedge-document.

Chapter 25

Bad Design—Our Biochemical Pathways: If Cats Don't Die of Scurvy, Then Why Do We?

Figure 25.1   Why is this cat smiling?
Because it can't get scurvy.

One of the things that ID people love to point at is our biochemical pathways. They say that these pathways are so complicated and so perfect that they must be the work of an intelligent Creator.

I say that if indeed the Creator did design our biochemical pathways, then he has to account for the millions of cases of scurvy and the resulting miserable human deaths that have happened on his watch.

Here's what a biochemical pathway is: a complicated set of chemical interactions that work together, usually in a particular sequence, to produce a particular substance that is needed by the body. These series of reactions are sometimes very long and complicated.

Here's why our biochemical pathways are a case of *bad design*: we have the pathway for making vitamin C, but it isn't finished. Having an incomplete pathway for making a vital nutrient is *bad design*.

When our bodies lack vitamin C, we get scurvy. When we get scurvy, our bodies can't produce the proteins that make up the tissues and ligaments that hold us together, and we literally fall apart, among other problems.

Scurvy is a really ugly disease, with brutal symptoms. Human beings can die of scurvy.

HERE IS WHAT HAPPENS WHEN YOU GET SCURVY: Blood vessels break and bleed under the skin, forming tiny red spots, and blood can ooze from our skin, especially around hair follicles. We get rough skin, internal bleeding, and blotchy bruises all over the body, particularly on the legs. Legs and arms can swell up, wounds will not heal, and old wounds that have been covered with scar tissue can reopen. We get weak, brittle bones. Skin can become dough-like or gangrenous, and gums can redden, recede, and blacken. Teeth loosen and fall out.

There is also growth cessation, tenderness to touch, weakness, bone fragility, and joint pain.

We get anemia from impaired iron absorption, frequent infections due to suppression of the immune system, muscle degeneration, pain, fatigue, hysteria, and depression.

Here are pictures of some of the symptoms of scurvy.

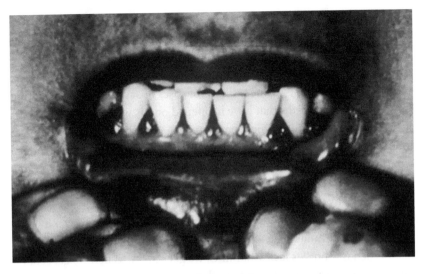

Figure 25.2   Symptoms of scurvy: gums.
Gums redden, recede, and blacken. Teeth loosen and fall out.

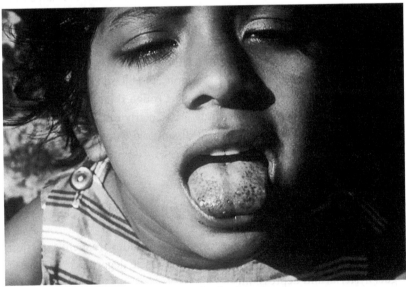

Figure 25.23   Scorbutic tongue.
Small blood vessels just under the surface of the tongue break and release blood.

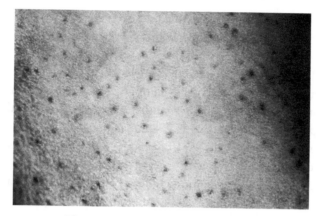

Figure 25.4 Symptoms of scurvy: skin.
Blood vessels break and bleed under the skin.

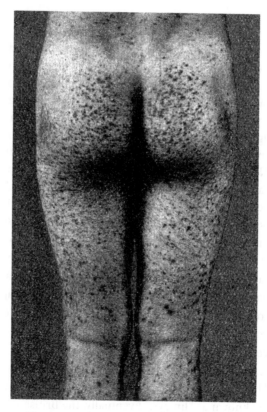

Figure 25.5 Symptoms of scurvy: legs, back.
We get rough skin and blotchy bruises all over the body, particularly on the legs. Bruises and exertion can result in internal bleeding. Legs and arms can swell up.

If we have scurvy for long enough, we die.

Having a biochemical pathway that isn't quite finished—and this can kill you—is *really bad design.*

So ID types love to talk about our biochemical pathways, but they don't mention the one for making vitamin C. That's because humans don't have a complete setup for making vitamin C—but cats do. I'll say that again: *cats can produce their own vitamin C —but humans can't.*

Does this sound like the work of an intelligent Creator? Does this Creator like cats better than humans?

Since we can't make vitamin C, human beings have to get that vitamin by eating certain fresh fruits and vegetables. Unfortunately, not everybody gets to have fresh fruits and vegetables.

THE BAD OLD DAYS

In fact, in the days before we had easy transportation and refrigeration, people in northern climates, often suffered from scurvy during the winter, when fresh fruits and vegetables were scarce. Indigenous people in Canada invented a tea made from the annedda tree to drink during the winter to prevent scurvy. Guess what the *annedda* is? It's the Eastern White Cedar. Does drinking a boiled infusion of this particular tree's wood and needles sound like a natural, or a delicious part of a diet? Do we think they drank this for fun? No. It was a concoction invented by people in a northern climate desperate for a way to prevent scurvy during the winter. Basically, we humans originally invented medicine in part to help save us from dying (too often) from ailments that could have been prevented if we'd been designed right.

People in other civilizations were not lucky enough to have heard of this cure, and they died in the winter of scurvy instead. People with scurvy are reported as early as in ancient Rome, and from as far away as China.

In the days of sailing ships, when there were long voyages and a lack of refrigeration, scurvy famously killed more sailors than drowning and piracy and all other diseases combined. A sailor on a voyage of exploration had a 50 percent chance of dying of some disease, usually scurvy, by the end of the voyage. Naval operations were limited by the amount of time that people could be at sea before they started to die from scurvy. It's why British sailors became known as "limeys." Eventually, the British government—to whom naval operations were extremely important—figured out that eating citrus fruits could prevent scurvy. So, every day, British sailors were required to drink a small portion of lime juice, so as to not get scurvy. The ship's cats were not required to do this, and they didn't get scurvy anyway.

In fact, cats don't eat fresh fruits and vegetables at all. I have never seen a cat sit down and eat an orange. Or even a cabbage. Yet they don't fall apart.

Figure 25.6   This cat didn't eat fresh produce. It didn't get scurvy.

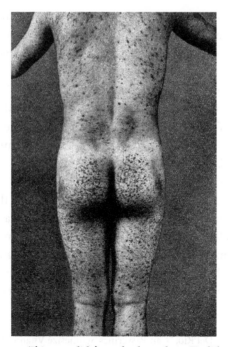

Figure 25.7   This man didn't eat fresh produce. He did get scurvy.

Is this any way for God to behave? It gets worse.

In Robert F. Scott's famous expedition to the South Pole on which he and all the men in his exploration party died after they reached the South Pole, all of the men in this party suffered from scurvy, which very likely contributed heavily to their deaths.

Meanwhile, there were seals lounging on the coast of Antarctica that *never* got scurvy. Do we think that the fish-eating seals ate fresh fruits and vegetables? Do we think that the seals ate fresh fruits and vegetables ever, in their entire lives? Do we think that the seals ate imported vitamin C tablets that hadn't been invented yet?

No—the seals made their own vitamin C.

What's more, on polar expeditions in general, the sled dogs and the men were away from fresh supplies for weeks and months on end. Yet the men got scurvy, and the dogs didn't. Why? Because the dogs made their own vitamin C.

Seals and dogs can make vitamin C but humans can't.

WHAT WAS THE CREATOR THINKING?

In fact *most* animals have the capacity to make vitamin C, but we don't. One of the few groups of animals that can't make their own vitamin C is us and our simian relatives. Simians are monkeys, apes, chimpanzees, humans, and so on.

Is this fair? Does the Creator like dogs better than us? What about cats and seals? For that matter, what about rats and vultures? They can make vitamin C, too. What have rats and vultures got that we haven't got? I'll tell you what they've got. They have a complete biochemical pathway for making vitamin C.[1]

And we don't.

It gets worse. We have the entire pathway except for the *last step*.

I'll say that again: We have the biochemical pathway that can make vitamin C, except that it's missing the *last step*. All we are missing is the ability to make one enzyme, in the *last step* of the process. Why are we missing this enzyme? One bad gene, that's all. We have one gene that mutated so we can't make one enzyme. This prevents us from making vitamin C.

---

1. Carlos Martinez Del Rio, "Can Passerines Synthesize Vitamin C?," *The Auk*, 1997, 114 (3), 513–16.

Having a biochemical pathway that's complete except for the last step is *bad design*.

And again, having a biochemical pathway that is complete except for the last step and this can kill you is *really bad design*.

But having a biochemical pathway that is complete except for the last step, and this can kill you, and the only thing you are missing is the ability to make *one enzyme* in the last step, and *nearly all other animals have the full and complete pathway*, is *excruciatingly bad design*.

Having a biochemical pathway that nearly all other animals have, but we only have part of it, because we are missing the ability to make one enzyme in the last step is not the sign of an intelligent Creator.

In fact, it's proof that we've evolved, rather than being designed.

EVOLUTIONARY THEORY AND THE SIGNATURE IN THE CELL

Why? Because evolutionary theory predicts this lousy biochemical pathway, and ID does not. In fact, ID wants to tell you that all of our pathways are wonderful. That's what the design inference is all about. Here's why this lousy pathway is predicted by evolution.

Most other animals can make their own vitamin C, by way of a specific biochemical pathway. Humans can't. But evolutionary theory says that we humans share common ancestors with the many species that can make vitamin C.

Evolutionary theory therefore says that we should find evidence of this shared ancestry, with at least some portion of the ancestral vitamin C-making pathway being found in humans.

And that's what we find. Humans and other simians have that vitamin C-creating pathway, but it's incomplete. It's missing the ability to create that one enzyme. Does this look like the work of an intelligent Creator? No. It looks like the work of a random, and in our case unfortunate, mutation.

Dr. Stephen Meyer of the Discovery Institute claims that the DNA in our cells shows the "signature" of our Creator.[2] Unfortunately for Dr. Meyer, that "signature" in our vitamin C pathway says *evolution*.

---

2. Stephen C. Meyer, *Signature in the Cell: DNA and the Evidence for Intelligent Design* (New York: HarperCollins, 2009), 470.

HOW DID WE GET THAT WAY?

Let me fill in the rest of the picture. Some ancestors of ours had a mutation that left us unable to manufacture that one last enzyme needed for making vitamin C. If we had been dogs or cats or other carnivores that get next to no vitamin C through their diets, then individuals with that mutation would have died out quickly. But because we simians have always eaten a mixed diet of meats plus fruits and vegetables, those ancestors with that unfortunate mutation survived, because the mutation didn't kill us before we reproduced . . . too often.

This would have continued to work well enough if humans hadn't turned out to be particularly successful primates that spread all over the world. Most nonhuman primates live in tropical and subtropical areas where there are plenty of fresh fruits and vegetables to eat throughout the year. With all that vitamin C being taken in through their diets, scurvy generally wasn't a lethal problem for our primate ancestors, even with their defective gene that wouldn't allow them to produce their own vitamin C.

But humans have spread all over the world, including settling into climates that didn't provide them with year-round fresh produce or other sources of vitamin C. So scurvy killed us from time to time. But it didn't kill the cats and rats and dogs living with us. Curious humans also set off on voyages of exploration in sailing ships where they had no access to fresh produce for months on end. Then scurvy killed a whole lot of them. But it didn't kill the ship's cats, or the ship's rats.

Now you may say that voyages of exploration are an unnatural environment, and that people in more normal conditions wouldn't get scurvy. But people in northern climates routinely got scurvy during the winter.

Why would the Creator make human beings suffer from scurvy during the winter, but not rats?

For your additional enjoyment, here is a description of sailors who were dying of scurvy in the 1830s. This is a passage from *Two Years Before The Mast*,[3] on the joys of suffering from scurvy during the age of sail. In this sequence, a young sailor is describing the effects of scurvy on members of the crew.

---

3. Richard Henry Dana, *Two Years Before The Mast* (New York: P. F. Collier and Son), 1909. This is an autobiographical account of a young American's experience as a hired sailor from 1834 to 1836.

The scurvy had begun to show itself on board. One man had it so badly as to be disabled and off duty, and the English lad, Ben, was in a dreadful state, and was daily growing worse. His legs swelled and pained him so that he could not walk; his flesh lost its elasticity, so that if it was pressed in, it would not return to its shape; and his gums swelled until he could not open his mouth. His breath, too, became very offensive; he lost all strength and spirit; could eat nothing; grew worse every day; and, in fact, unless something was done for him, would be a dead man in a week, at the rate at which he was sinking.

Aren't you glad that the Creator gave you the opportunity to die of such a horrible disease, but spared all the sweet little rats?

Figure 25.8   Rat.

Chapter 26

Why Does Intelligent Design Act so Much Like the Tobacco Lobby?

The tobacco lobby was a pioneer in denying scientific evidence. Once the link between smoking and cancer became clear in the 1950s, the tobacco industry stood to lose billions of dollars if people quit smoking for fear of getting cancer.

With billions of dollars at stake, the tobacco industry did everything it could think of to try and deny the connection between tobacco use and cancer. It needed to keep people thinking that smoking was safe, at least long enough for them to get addicted.

With the amount of money that it had and the amounts it stood to lose or gain, the tobacco lobby could afford to develop whole new realms of dishonesty, and it did this with great success. So much so that subsequent science-denying enterprises such as ID have simply copied their playbook.

Here are the tobacco lobby's ploys for denying science.

Manufacture a "Controversy"

The tobacco lobby is famous for manufacturing doubt. In the face of overwhelming evidence of the causal connection between smoking and cancer, the tobacco lobby continuously said that there was still uncertainty in the results. It also denied the fact that tobacco is addictive by insisting that scientific results on the subject were uncertain.

Here are a few gems from the 1954 advertisement by the Tobacco Industry Research Committee called "A Frank Statement to Cigarette Smokers":

> Recent reports on experiments with mice have given wide publicity to a *theory* that cigarette smoking is in some way linked with lung cancer in human beings." [Italics mine.]
>
>> Distinguished authorities point out:
>> 1. That medical research of recent years indicated many possible causes of lung cancer.
>> 2. That there is *no agreement among the authorities* regarding what the cause is.
>> 3. That there *is no proof* that cigarette smoking is one of the causes . . .

Doesn't this sound like ID's arguments?

Hire a Few People with Fancy Degrees to Be Your Public Face

They got a few scientists to be lobbyists. They paid them very well. These former scientists prostituted themselves in public by saying that the tobacco/cancer link was uncertain. One of the people they hired was a guy named Fred Singer. Even as late as 1993, Fred Singer was denying the cancer risks associated with secondhand smoke. Dr. Singer has a PhD in physics.

Start a Fake "Research Institute"

The tobacco lobby also came up with the Tobacco Research Institute, which didn't do much of any laboratory research, but which had nice stationery and other things that made it look like a real, respectable scientific research organization. It was just a front for tobacco's well-paid lobbyists, but the "Tobacco Research Institute" sounds very official, and makes it sound as though it's just another scientific research institute like, say, the Massachusetts Institute of Technology (MIT).

This way, tobacco lobbyists could pretend to be well-meaning, neutral academics, just trying to inform and protect the American people.

Pretend It's a Debate

From this platform they then went on to claim that because of the "un-certainty" (that they created), the two "sides" had equal merit, which they didn't. It also turned very serious cancer research into something that sounded like a high school debate.

The tobacco lobby further insisted that simply because it *claimed* that the two sides had equal merit, that meant that the two sides did have equal merit.

Repeat Untruths Unashamedly

The tobacco lobby happily repeated the same untrue things over and over, no matter how often they were corrected in public with facts that proved their assertions wrong.

Pretend You're a Victim

The tobacco lobby loved to pretend that they were victims, instead of admitting that they were a powerful multibillion-dollar industry. They pretended that smokers were victims of anti-smoking campaigns, rather than being the victims of a powerful and addictive substance that the tobacco lobby was making billions of dollars from selling. They objected strenuously to efforts to help people stop smoking, or get away from smoke-filled areas, but they did not object to their own multimillion-dollar advertising campaigns aimed at getting people to start or continue smoking. They also pretended that they were being victimized if people were warned about the proven health hazards of smoking.

People are generally kind and forgiving toward victims. So the tobacco lobbyists probably thought that by portraying themselves as victims, they could get people's sympathy, and then people would be less likely to hold them responsible for tobacco's deleterious health effects.

Talk About Freedom

The tobacco lobbyists loved to talk about freedom. They talked about how smokers "rights" were being infringed. Of course, when they talked about freedom, they didn't mean the freedom of the American people to know the true dangers of smoking cigarettes. That freedom they kept as far from the American people as possible and they spent a lot of money doing it. Instead, they insisted that informing the American public of the tobacco/cancer link was an unwarranted intrusion on their freedom.

The real reason that the tobacco lobbyists loved to talk about freedom was that it allowed them to change the subject. They could move the conversation away from all the deaths caused by smoking, and pretend instead that they were just sticking up for freedom, which is an emotionally charged topic. It allowed them to pretend that they were heroes instead of very well-paid lobbyists.

Hook 'Em While They're Young!

Tobacco companies made deliberate attempts to hook people on cigarettes while they were young. These attempts included everything from advertisements that featured kid-oriented cartoon characters to giving away free cigarettes to small children at playgrounds.[1]

So, to recap, the tobacco lobby's playbook looked like this:

1. Manufacture "uncertainty." It doesn't matter how you do it. You don't need any results of your own. Just call other people's work into question. Use the word "controversy" when describing the opposition's scientific results.

2. Get some spokespeople who have fancy degrees, who will say anything so long as you pay them enough.

3. Invent an "institute" as a front for your lobbyists.

4. Pretend it's a debate. Claim that "uncertainty" means that both sides have equal merit.

5. Repeat things that are untrue, even after you have been publicly corrected.

6. Portray yourself and your followers as victims. It gets people's sympathy.

7. Talk about freedom. It's very important to talk about freedom, since it's a way of changing the subject and moving the conversation away from the fact that tobacco products cause cancer.

1. Milton J. Valencia, "Smoker's suit wins award of $71m," *Boston Globe,* December 5, 2010.

8. When possible, convince children to be your followers—the younger, the better.

The tobacco lobby did all this hoping to cover up for the fact that all the experimental data from laboratories all over the world showed a clear link between tobacco products and cancer.

INTELLIGENT DESIGN FOLLOWS THE TOBACCO LOBBY'S EXAMPLE

The intelligent design lobby has done the same thing.

1. *They have manufactured "uncertainty" about evolution where none exists.* They claim that uncertainty exists, even though 99 percent of all scientists state that evolution by natural selection is a fact.

2. *They have some lobbyists with fancy degrees*, like Michael Behe and William Dembski, both of whom have PhDs. Also remember Dr. Fred Singer, who I mentioned earlier as denying the link between secondhand smoke and cancer? Now he's denying the link between CO_2 and global warming. You know where his coauthor gave a public speech on the topic? The Discovery Institute.

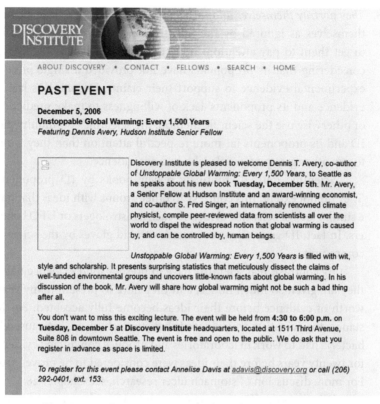

Figure 26.1   Event announcement from the Discovery Institute.
Here is the announcement for the event denying human-caused global warming. The event took place at the Discovery Institute. The coauthor of the book being presented is Fred Singer.

3. *They have a fancy-sounding institute, the "Discovery Institute,"* which has no scientific equipment, not even a microscope, but does have nice stationery.

4. *They claim that "uncertainty" means that ID, which has no experimental evidence, has just as much merit as evolution by natural selection, which has 150 years of scientific evidence behind it.*

5. *They say things that are untrue, even after they have been corrected.* For instance, they still claim that *Of Pandas and People* (renamed *Design for Life*) was neither a creationist nor a religious text. See Chapter 6 for more details.

6. *They portray themselves and their movement as victims.* They portray themselves as ignored by the scientific "establishment" and unable to get them to pay attention. In fact, ID is treated remarkably well considering that its proponents have not provided a single piece of experimental evidence to support their claims. Despite ID's lack of evidence and its proponents' lack of willingness to make predictions or otherwise use the scientific method, the scientific community pays ID and its proponents far more respectful attention than they would to any other group of people who had no evidence.

Scientific journals routinely review books by ID proponents. They do not extend this courtesy to other groups with ideas that have a similar lack of scientific evidence, such as astrologers or UFO believers. In fact, ID generally gets treated with kid gloves by the scientific community.

By contrast, real scientists have to present solid evidence before they are given so much notice, and they have to present many years' worth of evidence before their ideas become fully accepted. For instance, the scientists who showed that stomach ulcers are caused by bacteria had to work in laboratories and produce convincing evidence for twenty years before their ideas were considered to be proven true. For more discussion of stomach ulcer research, see Chapter 28.

What's more, ID proponents get interviewed on television and radio, which is more attention than most serious scientists ever get, and which does not make it sound as though intelligent design is being silenced. ID gets lots of funding from the Discovery Institute, and gets lots of political support from the likes of former and current presidential candidates Rick Santorum, Mike Huckabee, Rick Perry, Ben Carson, and Michele Bachmann.

7. *Then they talk about freedom.* This is a classic move to try and change the subject and move away from their very weak case, and toward an emotional subject that Americans care deeply about, but which is not relevant to whether or not ID has any scientific merit.

So ID promoters always try to bring up the phrase "academic freedom," even though public school teachers don't have it and have never had it. It's just a way for them to try and change the subject to something emotional, so that people won't notice how bad their arguments are in terms of solid evidence. I talk more about academic freedom and what it is, and is not, in Chapter 16.

The ID lobby does all this, hoping to cover up for the fact that there is no experimental evidence in favor of ID.

8. *They're deliberately targeting young people.* The ID lobby wants to promote ID to school children, in public schools, using the children's authority figures, teachers, as their pitchmen.

Straight to Textbooks

Then they do the tobacco lobby one better, by insisting that their idea should not only be taught in schools, it should be *printed in textbooks before it has produced any experimental evidence.*

This is in direct contradiction to how real science is done. When real scientists have a new idea, they make predictions, do experiments, and get their results published in scientific journals that are vetted and reviewed by other scientists. This goes on for many years before the new idea is fully accepted. Only then does it go into textbooks. But all this scientific work and waiting is just too tedious for ID promoters. They want their ideas taught as truth, or perhaps I should say gospel, right now.

Imagine what would have happened if science textbooks had been required to print the bogus claims of the cigarette manufacturers.

Now that I've talked about the folks in the tobacco and ID lobbies, let's move on to some much nicer creatures. Let's talk about sharks.

Chapter 27

Bad Design—Our Teeth, or, Why Is This Animal Smiling?

Figure 27.1 Q: Why is this shark smiling?
A: Because it doesn't have to go to the dentist.

TEETH

Getting back to *bad design*, think about our teeth. We humans only get two sets of teeth in our entire lifetimes. First we get our baby teeth, which drop out during childhood. Then we get our adult teeth. We try to keep them for the rest of our lives, because if they rot out, we have no more teeth. By age 70, we've spent 91 percent of our chewing careers with our adult teeth—if we manage to keep them that long.

Here's the problem: teeth come into constant contact with food. Unfortunately, food for us is also food for the bacteria in our mouths, which can infect our teeth and gums. As a result, tooth and gum infections are pretty common. In animals that have the same set of teeth for their entire adult lives, it's pretty easy and common for long-term dental problems to set in, which can easily lead to dangerous infections. Even in animals such as dogs or cats, which only live for a decade or two, serious and even potentially fatal tooth and gum infections are fairly common.

Basically, there are bacteria that stick to the surfaces of your teeth. They form a sticky matrix that traps food particles, which creates plaque deposits. These both feed and protect bacteria. The bacteria living in the plaque deposits create acids that erode your teeth. This produces cavities. If bacteria erode their way all the way to the centers of your teeth where the soft pulp is, then the pulp gets infected. This is called pulpitis. The only way to fix this is to get a root canal. Otherwise, you get an infection in your jaw. Back on the surface, bacteria can also attack the connection between your teeth and gums. This causes your gums to recede, and then the bacteria can attack lower down on your teeth. This can eventually cause your teeth to fall out. What's more, anytime you have an infected tooth or gum, or a hole from a cavity, it's a highway for bacteria to enter your body and attack the bone of your jaw, your brain, your heart, and elsewhere. This can lead to death. Dental problems are very nasty when not dealt with by modern dentistry.

Having a structure that invites painful and deadly infections is *bad design*.

Humans have the potential to live for many decades, but dental problems can often cut life short in the natural environment. Having a structure that routinely ends lives decades early is *bad design*.

That's why we invented dentists.

Here are some diagrams of how our teeth look when they are well, and how they look when they are infected. Keep in mind that the very act of their existing and doing the job that teeth are supposed to do puts them into contact with the bacteria that kill them, and *feeds the bacteria that kill them*. Having a structure that feeds the bacteria that will then attack and kill it is *bad design*. Having a structure that makes it easy for those well-fed bacteria to then attack the rest of the body is *really bad design*.

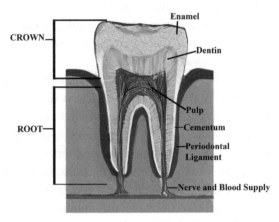

Figure 27.2   A diagram of a healthy tooth.

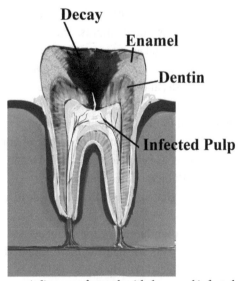

Figure 27.3   A diagram of a tooth with decay and infected pulp.
Here is a diagram of a tooth in which bacteria have gotten into the soft inner core of the tooth. Notice how this infected core leads right to your blood vessels—which lead to the rest of your body! No wonder people with bad teeth die young.

So, unfortunately, our teeth and gums get attacked by bacteria on a regular basis. In the old days, infected teeth and gums were a routine part of life—and death.

Every single infected tooth was a threat to life if the infection spread. This was before the days of antibiotics, so an infection that spread to the rest of the body could easily kill you. Infected gums often caused the face to swell tremendously. People were weakened and sickened, and frequently died. Having an important part of the body that wears out early, and invites deadly infections, is *bad design*.

THE PAIN OF TOOTH EXTRACTION WITHOUT ANESTHESIA WAS BETTER THAN LIVING WITH INFECTED TEETH

Even if people lived with an infected tooth, they were in great pain. So great, in fact, that having the tooth pulled seemed like a good idea by comparison. I'll say that again: the threat to life and the pain of infected teeth was so great that having teeth pulled was a far better idea—*and this was in the days before the invention of anesthesia*. So people sat in a dentist's chair and had the tooth pulled manually with no painkillers in order to live another day. All because our teeth rot and don't replace themselves.

This is not entirely due to a modern diet. People living in the natural environment, prior to the invention of agriculture, generally had a life expectancy of about thirty years.[1] These early deaths were attributable to many causes, but infected teeth were one source of mortality. What's more, people living in the natural environment get tooth infections and rotten teeth on a regular basis. If there are no dentists, this will shorten their lives.

In the old days, people lost many teeth in the course of a lifetime. Many lost all of them. This is why false teeth were invented. Most people from middle age onward had false teeth until fairly recently. Losing teeth is awful, and the pain of extraction is excruciating, but it was better than dying.

So, one reason that humans are able to live long lives today is because we invented dentistry—not because we are designed well.

Wouldn't it be nice if we could just replace rotten teeth with good ones? How hard would that be? If we just grew new teeth on a regular basis, then all these dental problems would go away. Cavities wouldn't have time to grow. Dental infections wouldn't have time to set in and cause other damage, and we could live long, natural, healthy lives *without dental bills*.

---

1. Oded Galor and Omer Moav, "The Neolithic Revolution and Contemporary Variations in Life Expectancy," *Brown University Working Paper*, August 28, 2007, http://www.brown.edu/Departments/Economics/Papers/2007/2007-14_paper.pdf.

Yet sharks can live for many decades. Some live more than 100 years. And they never visit the dentist. Their teeth are fine. Why? Because sharks continuously replace their teeth throughout their lives. Great white sharks replace all their teeth about every 230 days. Lemon sharks replace their teeth every eight to ten days. Those teeth don't have time to get cavities! No root canals for these fish! No tooth extractions, no braces, no dental bills.

It seems like such a simple fix!

So, does the Creator like sharks better than people?

Figure 27.4   Why does this animal deserve a tooth pain-free life?

However, sharks without dental problems may be a good thing after all, since the only thing I can think of that's scarier than a great white shark is a great white shark with a dentist's drill.

Figure 27.5   "This won't hurt a bit."

Actually, I can think of something scarier, and that's a megalodon with a toothache. Megalodon (*C. megalodon*) were sharks that existed about 23.8 to 5.3 million years ago, back in the Miocene period.

They were probably the biggest, baddest sharks ever. They were about forty feet long or longer at their biggest, which makes them larger than any fish alive today, and twice the size of the largest great white shark. They had teeth the size of grapefruits, only much sharper.

Figure 27.6   The size of a *megalodon* tooth is frightening.

Fortunately for them, the Creator decided to not let the deadly megalodons have toothaches, so they were spared all that pain. Too bad humans didn't get the same deal.

Figure 27.7   So is the size of a *megalodon* jaw.

So if the Creator really does like sharks better than people, what does this say about his personality?

Chapter 28

The Discovery Institute Hasn't Discovered Anything (Sort of Like the Tobacco Institute)

"...we are convinced that in order to defeat materialism, we must cut it off at its source. That source is scientific materialism."

— *THE WEDGE STRATEGY*

The Discovery Institute is an oddly named organization, especially since it hasn't discovered anything. In fact, it doesn't even own a microscope. Sort of like the way that the Tobacco Research Institute didn't do any scientific research.

What they are good at doing is complaining about other people's scientific research. However, they are very bad at figuring out what they themselves actually think.

They are also very good at arguing. In fact, that's mostly what they do. Except that they call this arguing "research." This is different from the research that scientists do, which generally involves going into the laboratory, or doing field work, and writing down observations, and doing experiments, and subjecting their results to statistical analysis, and so on. The only thing that the folks at the Discovery Institute do is argue. A lot. They argue on television, they argue on the radio, they argue in print. They never talk about their experiments.

Q: WHEN IS AN INSTITUTE NOT A SCIENTIFIC RESEARCH INSTITUTE?

A: WHEN THEY DON'T DO SCIENTIFIC RESEARCH.

Papers written by people at the Discovery Institute contain lots of arguments. Scientific papers, by contrast, tend to be fairly dry, and they stick to facts such as how the experiments were done, and what the results were. Arguing is rarely a part of a scientific paper. Scientists tend to believe that the facts speak for themselves. This is why it is easy to tell a scientific paper from almost anything else. Scientific papers talk almost exclusively about experiments, observations, data, results, and other things that matter to scientific research. What you will not find in scientific papers are thundering arguments, fiercely held contentions, or other forms of rhetoric commonly used by theologians and philosophers.

In fact, the contrast between scientific papers and other forms of writing is so pronounced that I did an experiment to check the truth of the claim that ID really is science.

Here's what I did:

Articles by design proponents that were claimed to be both scientific and peer reviewed and that were available on the Discovery Institute (DI) website were analyzed, looking for the word *data*, including both *data* and *database*, which would indicate a quantitative approach to their work. I also looked for words containing the root "*argu-*," which would indicate a debate-like approach to scholarship more characteristic of philosophy or religion. These results were compared to word counts from an identical analysis of papers by evolutionary biologists at the Smithsonian Tropical Research Institute (STRI).

I predicted that articles by ID proponents from the Discovery Institute would use words with the root *argu-* more than the word *data*, while articles by scientists from the STRI would have the reverse proportions. What I found was that the root *argu-* was frequent in the papers from the Discovery Institute but nearly nonexistent in the papers from the Smithsonian

Tropical Research Institute. By contrast, *data* was abundant in the STRI papers, but infrequent in papers from the DI (see figure 28.1).

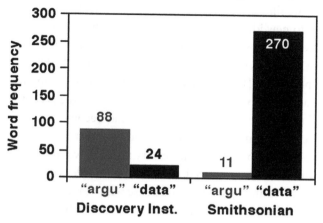

Figure 28.1   Histogram of word frequencies.
The frequency of the use of the words *argu-* and *data* by proponents of intelligent design at the Discovery Institute, and by scientists at the Smithsonian Tropical Research Institute (STRI). Scientists at STRI made heavy use of data, but rarely used or cited argumentation in their articles. Discovery Institute writers rarely referred to data, and never to testing a hypothesis, but referred heavily to argumentation. The difference was extreme. These results were significant to $p < 1.9 \times 10^{-53}$. That's: significant to $p < .00000000000000000000000000$ 00000000000000000000000019, using a chi-square test.

Having ascertained that Discovery Institute authors rarely use the word *data*, I then examined every case in which they did, to see if they ever used it to refer to their own testing of a prediction that they had made, using a hypothesis generated by ID. I found that out of the twenty-four instances in DI-published papers in which the word *data* was used, nineteen referred to data generated by other people, usually data on Cambrian fossils; four referred to data as a concept; and one referred to data that were original, but that did not test a hypothesis.

In short, in all of the supposedly peer-reviewed ID literature that the Discovery Institute has published, there was not a single instance of hypothesis testing.

So although proponents of ID like to claim that it is science, these data, taken from their own writings, strongly contradict this claim.

These results were published in the September 2015 issue of *The American Biology Teacher*, a peer-reviewed journal.

The Discovery Institute has had twenty years in which to come up with hard scientific data in support of ID. They have not done so.

Real, significant science can be done in twenty years. Scientists are very open to new ideas, when they are backed up with solid experimental data. Since the 1970s, other scientific paradigms have been upended and replaced with newer, better ones, when the newer ones had better evidence. For example, up until the 1970s, it was considered to be common knowledge that stomach ulcers were caused by stress. This was the accepted explanation in all medical textbooks. Since then, it has become accepted common knowledge that stomach ulcers are caused by infection by the bacterium *Helicobacter pylori*, because the necessary research was done and convincing scientific evidence produced, which led to this idea's acceptance. So the scientific community is willing to dispense with old ideas, however venerated, in the face of convincing new evidence. During this same time period, ID has not produced convincing new evidence in favor of its claims.

WHAT WOULD A REAL THEORY OF ID LOOK LIKE?

If the Discovery Institute were serious about science, it would put together a real, coherent theory of intelligent design, and then do experiments that might provide evidence for it.

A real, scientifically serious theory about ID would look different from the stuff that ID promoters talk about. It would include research on how often the Creator comes around. How we can tell newly designed species from slight variations in old ones. When should we next expect to see the Creator, and what would be on his to-do list?

Instead, the ID folks disagree among themselves about just about everything. They cannot agree on the age of the earth, what fossils are, whether or not ID is religion, or whether the Creator worked only once or has shown up many times. They cannot make coherent predictions about the natural world.

In fact, I'm not sure that they could agree among themselves that gravity makes things fall down.

A genuine theory of ID would also make testable predictions. Without testable predictions, you don't have a theory. A genuine theory of ID would also include genuine experimental results.

WHO CAN SUCCESSFULLY CONTROL ANTIBIOTIC RESISTANCE?

Antibiotic and antimicrobial resistant organisms present a golden opportunity for ID to prove itself. ID proponents could successfully provide evidence that ID exists by predicting how to control antibiotic resistant bacteria in a way that is different from the ways predicted by evolution by natural selection.

If ID promoters then do the experiments in the lab showing how they can control antimicrobial resistance using predictions made exclusively by ID, then they will have convincingly shown evidence for their case.

What we have found thus far, however, is that *evolution by natural selection both predicts these problems, and shows us ways of dealing with them. ID has done neither.*

In fact, in the more than twenty years that ID has been in existence, none of its promoters has run a single experiment. The only thing they've run is their mouths. They've had lots of money, but they still haven't produced a single experimental result. Remember, the Discovery Institute doesn't even own a microscope.

During this time, evolutionary biology has racked up hundreds *more* papers showing good evidence for evolution by natural selection. In fact, evolution now has 150 years of scientific evidence to back it up. What's more, the theory of evolution by natural selection has led to medical breakthroughs.

EVOLUTIONARY THEORY LEADS TO MEDICAL BREAKTHROUGHS

Scientific breakthroughs have resulted from accepting that evolution by natural selection is true. Evolution by natural selection has helped researchers find new treatments for cancer.[1]

Evolution by natural selection has helped researchers discover a new hormone. In fact, "Darwin led us to this new hormone," said Aaron Hsueh, PhD, an endocrinologist and professor of obstetrics and gynecology, whose laboratory discovered an important appetite-suppressing hormone. His

---

1. L. M. Merlo et al., "Cancer as an Evolutionary and Ecological Process," *Nature Reviews Cancer*, 2006, 6 (12), 924–35.

research was sponsored by Johnson & Johnson Pharmaceutical Research and Development, LLC.[2]

ID has not helped in either cancer research or hormone research.

FOLLOW THE MONEY

If ID is so right, then why don't big drug companies and big agricultural companies do research that's based on it?

Follow the money. Pharmaceutical and agricultural research companies solve problems and make money when they make predictions using evolutionary biology. They risk their money on research based on evolution by natural selection. They don't risk their money on ID. Why? Because it's not science. It doesn't make predictions. Its only product is uncertainty. Medical breakthroughs based on ID do not exist.

Pharmaceutical and agricultural companies are in business to make a profit, and will invest in just about anything that they think will work. They are not part of a government plot, an atheist plot, or anything else. They just want good, usable results, and they will use whatever methods produce them. ID doesn't produce them. Drug and agriculture companies are in business to make money, and the smart money is behind evolution-based, real science.

Always remember: the only thing that ID produces is uncertainty. Who do you want to go with, ID promoters who only manufacture "uncertainty," or the evolutionary biologists, who get real, useful results?

2. "Stanford Scientists' Discovery of Hormone Offers Hope for Obesity Drug," Stanford University School of Medicine, news release, November 10, 2005.

Chapter 29

The Publishing Scandal that Rocked the Discovery Institute

One odd thing about the Discovery Institute is the way they publish their "scientific" papers in religious journals. I discussed this in Chapter 6. But did you know that they once got a paper published in a genuine biological journal? Unfortunately for the DI, what should have been a moment of triumph became a farce and a long-lasting scandal.

Here's the story.

PROCEEDINGS OF THE BIOLOGICAL SOCIETY OF WASHINGTON

On August 4th, 2004, a paper called "The origin of biological information and the higher taxonomic categories" was published in the *Proceedings of the Biological Society of Washington*, a genuine biological journal. The paper was by Stephen Meyer, who works for the Discovery Institute and is a big ID promoter. The paper promotes ID. It argues a great deal, and presents no new data. Why was the paper published there? Basically, because it was the only journal that the Discovery Institute could get it into. And why could they get the paper in there and nowhere else? Well, maybe it's because the editor at the time now gets money from Discovery Institute.

The editor's name is Richard Von Sternberg. He was the one single person at the Biological Society of Washington who was in charge of which papers could get published, and which papers couldn't. "But what about peer review?" I hear you cry. Well, Mr. Von Sternberg says that he used peer review on this paper, but he won't tell who the reviewers were, because it's a secret.

"But how do you know he's being paid by the Discovery Institute?" you ask. Because we have their tax returns, that's how. Look! After he published the Discovery Institute's paper, Von Sternberg wound up on their payroll! How remarkable! Do you think it is a coincidence?

Here's the Discovery Institute's tax return for 2007. In 2007, Mr. Von Sternberg got $77,375 in pay from the Discovery Institute, and another $16,581 in expenses.

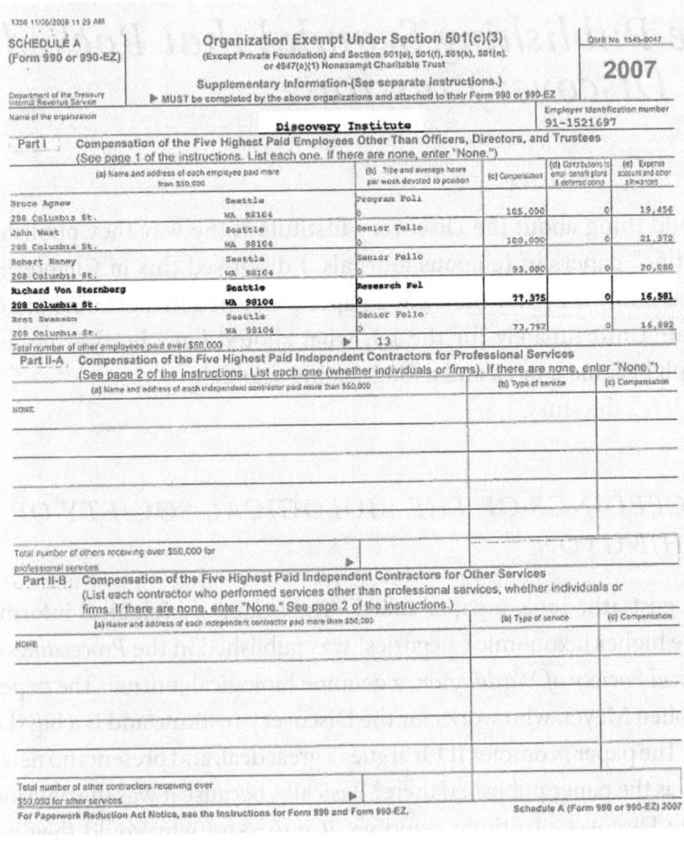

1356 11/06/2008 11 29 AM

SCHEDULE A (Form 990 or 990-EZ)
Department of the Treasury Internal Revenue Service

Organization Exempt Under Section 501(c)(3)
(Except Private Foundation) and Section 501(e), 501(f), 501(k), 501(n), or 4947(a)(1) Nonexempt Charitable Trust
Supplementary Information-(See separate instructions.)
▶ MUST be completed by the above organizations and attached to their Form 990 or 990-EZ

OMB No 1545-0047
2007

Name of the organization: Discovery Institute
Employer identification number: 91-1521697

Part I Compensation of the Five Highest Paid Employees Other Than Officers, Directors, and Trustees (See page 1 of the instructions. List each one. If there are none, enter "None.")

(a) Name and address of each employee paid more than $50,000		(b) Title and average hours per week devoted to position	(c) Compensation	(d) Contributions to empl. benefit plans & deferred comp	(e) Expense account and other allowances
Bruce Agnew 208 Columbia St.	Seattle WA 98104	Program Poli 0	105,000	0	19,456
John West 208 Columbia St.	Seattle WA 98104	Senior Fello 0	100,000	0	21,372
Robert Nanny 208 Columbia St.	Seattle WA 98104	Senior Fello 0	93,000	0	20,080
Richard Von Sternberg 208 Columbia St.	**Seattle WA 98104**	**Research Fel 0**	**77,375**	**0**	**16,581**
Bret Swanson 208 Columbia St.	Seattle WA 98104	Senior Fello 0	73,792	0	16,882
Total number of other employees paid over $50,000 ▶		13			

Part II-A Compensation of the Five Highest Paid Independent Contractors for Professional Services (See page 2 of the instructions. List each one (whether individuals or firms). If there are none, enter "None.")

(a) Name and address of each independent contractor paid more than $50,000	(b) Type of service	(c) Compensation
NONE		
Total number of others receiving over $50,000 for professional services ▶		

Part II-B Compensation of the Five Highest Paid Independent Contractors for Other Services (List each contractor who performed services other than professional services, whether individuals or firms. If there are none, enter "None." See page 2 of the instructions.)

(a) Name and address of each independent contractor paid more than $50,000	(b) Type of service	(c) Compensation
NONE		
Total number of other contractors receiving over $50,000 for other services ▶		

For Paperwork Reduction Act Notice, see the Instructions for Form 990 and Form 990-EZ. Schedule A (Form 990 or 990-EZ) 2007

Figure 29.1 A page of the 2007 tax forms from the Discovery Institute.
Showing Richard Von Sternberg's salary and expenses at the Discovery Institute.

During that same year, Stephen Meyer, the author of the article that Mr. Von Sternberg single-handedly accepted, received $102,500 in compensation and another $33,761 in expenses.

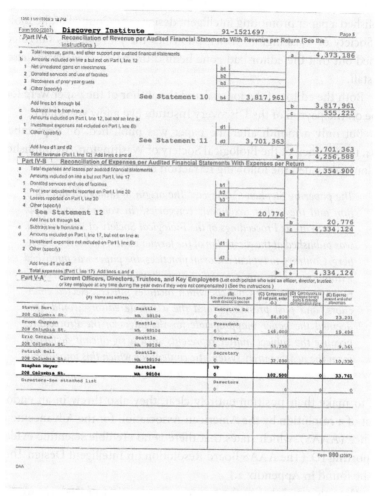

Figure 29.2   A page of the 2007 tax forms from the Discovery Institute.
Showing Stephen Meyer's salary and expenses at the Discovery Institute.

PROCEEDINGS OF THE BIOLOGICAL SOCIETY OF WASHINGTON—THE RETRACTION

On August 4, 2004, the *Proceedings of the Biological Society of Washington* published a paper promoting intelligent design. The Council of the Biological Society of Washington, which publishes the journal, was horrified. It turned out that the editor had gone behind the backs of everyone else on the staff.

Both the editor of the journal and the author of the paper were found to be on the payroll of the Discovery Institute just a few years later.

But only a month after the paper was published, on September 7, 2004, the Council of the Biological Society of Washington, the publisher of the journal, issued the following retraction of that article:

> *The paper by Stephen C. Meyer, "The origin of biological information and the higher taxonomic categories," in vol. 117, no. 2, pp. 213–239 of the Proceedings of the Biological Society of Washington, was published at the discretion of the former editor, Richard v. Sternberg. Contrary to typical editorial practices, the paper was published without review by any associate editor; Sternberg handled the entire review process. The Council, which includes officers, elected councilors, and past presidents, and the associate editors would have deemed the paper inappropriate for the pages of the Proceedings because the subject matter represents such a significant departure from the nearly purely systematic content for which this journal has been known throughout its 122-year history.*

Just to make their position entirely clear, they also threw in an endorsement of a resolution by the American Association for the Advancement of Science (AAAS), which states that there is no credible scientific evidence supporting ID. (The AAAS Board Resolution on Intelligent Design Theory can be found in Appendix 2.)

They also said that in the future, proper review procedures would be observed—that is, they are admitting that proper review did not take place in the ID paper's case.

PROBABILITY

ID folks often like to talk about probability. They say that getting the organisms we have today by way of random mutations followed by natural selection is too improbable to be true. They point out that the odds of getting any particular organism are exceedingly small, and therefore they claim that evolution is impossible.

This is like saying that because the odds of your being born are exceedingly small, given that your two parents had to meet and reproduce on just the right day to have that particular sperm and egg meet that later became you, and given that the same thing had to occur for all the generations before you, these odds are extremely small, and this somehow means that it is actually impossible that you were born at all.

It overlooks a number of important things. Mostly, it overlooks the fact that even if you yourself hadn't been born, people similar to you would have been born. Likewise, if today's organisms hadn't evolved through natural selection, other similar organisms would have. The odds of that are extremely high.

I point this out at this juncture because the really improbable event here is the idea that Stephen Meyer's paper could have been published in the *Proceedings of the Biological Society of Washington* without having a friendly relationship with the editor of the journal. That's a really improbable event. The entire rest of the journal's staff and publishers were appalled by the article. The journal in the past had stuck to systematics papers anyway, that is, papers that are about how organisms are related to one another. A theory paper with no biological evidence would have been rejected on that basis alone, whether or not it had been about ID. Only one editor, Mr. Von Sternberg himself, was in favor of it, and that one editor somehow slipped it past everybody else, by keeping it a secret from everybody else until after publication.

So, what are the odds that Richard Von Sternberg would have gotten a well-paid position at the Discovery Institute if he hadn't published the Discovery Institute's one and only article that's in a real biological journal? Even if it was retracted only a month later?

For that matter, what are the odds of that paper being published, if the editor hadn't had a very friendly relationship with the Discovery Institute?

Now once again, having talked at length about unsavory humans, I will get back to talking about much nicer creatures. I will tell you more about sharks.

Chapter 30

Bad Design—Sharks Get More Reproductive Options, Too

Sharks are famous for their poor social skills. In fact, they sometimes eat each other.

Figure 30.1   TV news footage of a great white shark savaged by another shark,
probably another great white.
Some sharks have poor social skills.

They are so aggressive that they will even bite people before they are born. I'm not joking. A marine biologist was once inspecting a pregnant female shark, and had his hand in her uterus. One of the *unborn pups* bit the marine biologist on the thumb.[1]

Now that's aggression.

It gets worse. Sharks are so aggressive that they will not only eat other sharks after they are born, they are so aggressive that they will eat each other *before* they are born.

Sand tiger sharks (*Carcharias taurus*),[2] also known as grey nurse sharks (*Carcharias taurus*),[3] have pups so aggressive that they eat their siblings while in the womb. By the time they are born, there is only one pup per womb.

Sharks also tend to have solitary lifestyles. This means that sharks can sometimes have a hard time finding a mate. You would too under the circumstances.

So sometimes, sharks have to get creative if they're going to reproduce.

One thing that sharks have got going for them, that we haven't, is that *female sharks can breed without males.*[4, 5] That's right. If no male sharks show up that the female wants to breed with, she can reproduce all by herself!

Think of the convenience! If no good males are available, a female shark can create young all by herself. If they are, she can reproduce in the usual sexual way. Think of how many human women suffer because of a lack of good mates, or suffer at the hands of bad mates. Sharks don't have to put up with this, and they can still reproduce without one.

So sharks find it easier to breed than humans do. Maybe the Creator is a shark after all.

1. George H. Burgess, "Shark Conservation in the Western North Atlantic: A Perspective," Florida Museum of Natural History, http://www.flmnh.ufl.edu/fish/organizations/ssg/regions/region12/conservation.htm.

2. Demian D. Chapman et al., "The behavioural and genetic mating system of the sand tiger shark, *Carcharias taurus*, an intrauterine cannibal," *Biology Letters*, 9 (3), June 23, 2013, DOI 10.1098/rsbl.w013.0003.

3. John Platt, "Artificial uterus could save grey nurse shark from extinction," http://blogs.scientificamerican.com/extinction-countdown/2009/02/18/artificial-uterus-could-save-grey-nurse-shark-from-extinction/.

4. Demian D. Chapman et al. "Parthenogenesis in a large-bodied requiem shark, the blacktip *Carcharhinus limbatus*." *Journal of Fish Biology*, 2008, 73 (6), 1473, DOI 10.1111/j.1095–8649.2008.02018.x.

5. Stony Brook University, "'Virgin Birth' By Shark Confirmed: Second Case Ever," *ScienceDaily*, http://www.sciencedaily.com/releases/2008/10/081010173054.htm#.

Chapter 31

Exploding the Cambrian Explosion

Dr. Dembski argues that the very existence of the Cambrian explosion is a sign that the intelligent designer is real. The Cambrian explosion is a very interesting thing. Basically, during the Cambrian geological period, lots of multicellular organisms came into being. Dr. Dembski claims that so many came into view so fast that they couldn't possibly have had time to evolve, so they must have been put together by a Designer.

However, Dr. Dembski has some problems. Let's start with the obvious. Mammals, birds, fish, amphibians, and reptiles are all post-Cambrian. Meanwhile, most Cambrian organisms are extinct. So what was the Designer doing with all those Cambrian animals—rehearsing?

More obvious stuff: the Cambrian period spanned from 543 to 490 million years ago. So the Cambrian period lasted 53 million years. A lot of evolution can take place in 53 million years.

What's more, multicellular organisms existed before the Cambrian period. We have fossils of multicellular animals that are 560 million years old, that is, they showed up 17 million years before the Cambrian period. These organisms are known as the Vendian fauna (named after the geological period in which they formed), or the Ediacaran fauna (named after the hills in Australia where many of these fossils were found).

Here is a fossil from the Vendian/Ediacaran Period. This creature is named *Dickinsonia costata.* There are many Ediacaran fossils that show many-celled organisms that were architecturally complex. They all appeared prior to the Cambrian period.

Figure 31.1   An Ediacaran fossil called *Dickinsonia costata*.

Darwin hadn't heard of pre-Cambrian fossils, because they hadn't been discovered yet. The field of fossil exploration was only just beginning back in the 1800s, and even dinosaurs themselves were not described until the 1820s.

We've learned a few things since Darwin published *The Origin of Species* in 1859.

Dr. Dembski, on the other hand, has no excuse for not having heard of pre-Cambrian organisms.

Other interesting things also happened during the pre-Cambrian times. For instance, continents became stable, and oxygen became plentiful in our atmosphere. The oxygen was the result of photosynthetic bacteria and simple plants being able to make oxygen through photosynthesis. Before oxygen was plentiful, it was impossible for multicellular animals to exist.

So multicellular animals got going during pre-Cambrian times, and then took off as the environment became more stable and more oxygen rich.

It is also interesting that Dr. Dembski bases his arguments on the fossil record. Yet ID proponents often claim that the fossil record is inaccurate, or even downright false. Again, it appears that these guys couldn't agree among themselves that gravity makes things fall down. I wish these guys would agree upon what their theory that they are so intent on pushing on the rest of us is.

Fish, Amphibians, Reptiles, Birds, and Mammals Are All Post-Cambrian.

We would not recognize most of the Cambrian organisms. They are not the modern creatures that we see today. If they were intelligently designed, why are they no longer with us?

From an evolutionary standpoint, the Cambrian creatures make sense. In them we see the first appearances of some of the features, and some of the groups, that later evolved into current-day organisms—but not all of the features and not all of the groups that are with us in the present day. Some of those Cambrian creatures are roots in our biological family tree. Researchers are now connecting these Cambrian groups using both fossils and data from developmental biology.

Chapter 32

Why William Dembski's "Information Theory" Isn't Very Informative

INFORMATION THEORY AND THE HUMAN GENOME

Dr. Dembski says that he uses information theory to prove that the human genome must have been designed, and by an intelligent designer, at that. He wrote about it in a book titled *The Design Revolution*. Unfortunately for Dr. Dembski, he doesn't use information theory, nor does he use the human genome.

Dr. Dembski claims that the human genetic code is information, sort of like the bits in a computer code. He then claims that this entitles him to use information theory on the human genome. This shows that he doesn't understand what an analogy is. The other problem is that after he says he's going to do information theory, he doesn't do information theory.

Information theory is a branch of applied mathematics. Mathematics relies on equations, formulas, theorems, and lemmas. Without equations, formulas, theorems, or lemmas, it's neither mathematics nor information theory. Dr. Dembski uses no equations, formulas, theorems, or lemmas. He has a whole book that says that it's about information theory, but there are no equations, formulas, theorems, or lemmas in it. This is like claiming that there's a forest when there aren't any trees. Imagine! An entire book claiming to present advances in a field of mathematics, but it contains no equations! Or formulas! Or theorems! Or lemmas! This is the first time in history that someone has pretended to advance a field in mathematics, without writing any actual mathematics. This is ridiculous. So Dr. Dembski doesn't so much *do* math as talk about how he *would do* math if only he bothered to write equations, formulas, theorems, or lemmas. He does talk about God, though, specifically the Christian God. In a book about mathematics, in which he doesn't write equations, formulas, theorems, or lemmas.

Dr. Dembski also doesn't use the human genome. The entire human genome was published in 2001, and everybody was allowed to see it. Yet in his 334-page long book, he doesn't once look at the real, biological human genome. What's more, he didn't look at all the research that's been done on the real human genome since 2001.

So after saying that he's going to do math, he doesn't do math. And after he says he's going to talk about the human genome, he doesn't talk about the human genome. Not the actual, biological human genome, anyway. In other words, he doesn't do what he says he's going to do, and he doesn't do it to the thing he says he's going to do it to. I guess he figures you just won't notice this if he just uses enough words, and fills a book with them. His book actually says very little, but is something of a masterpiece in saying very little but taking a great many pages to do so.

SCIENCE IS BASED ON REALITY

It has been said that years ago, engineers did a bunch of calculations, and "proved" that bumblebees can't fly. But bumblebees kept flying, thoroughly ignoring all the engineers who said they couldn't. That's the difference between science and everything else. A scientist will tell you that if bumblebees keep flying, then the engineers are wrong, and not the bumblebees.

Information theory, being a branch of applied mathematics, isn't science. It is an interesting field, but not science. Physics is science. Chemistry is science. Biology is science. The sciences find things out about the real, natural, material world by making observations and doing experiments. Information theory doesn't do these things, so it isn't science.

Information theory may be applied to scientific fields, but you can't prove something in science using only information theory. If information theory, when applied to a field, makes testable predictions, then those predictions can be tested experimentally. In this way, science can make use of information theory, just as science can make use of pocket calculators, but pocket calculators alone can't prove something in science in the absence of actual data. Likewise, information theory alone can't prove anything in science, in the absence of actual scientific data.

None of this stops Dr. Dembski. In fact, very little seems to stop Dr. Dembski, least of all reality. Mostly, Dr. Dembski seems to like writing, and producing large volumes of words appears to be more important than making any sense. Dr. Dembski likes using big words like "propaedeutic," which means introductory, when he could just say "introductory." He quotes Schopenhauer—but he but doesn't say what work he's quoting from.[1] He holds degrees in both mathematics and philosophy, neither of which are sciences. He's hasn't done a single experiment. What he appears to like doing is taking a field like information theory, which most people don't understand, and taking a field like evolutionary biology, which most people don't understand, and then combining them in a book so that nobody will understand it. I think he figures that if he just makes things seem complicated enough, then people will give up and simply agree that he's right, because they don't understand what he's saying.

INFORMATION THEORY AND EVOLUTIONARY ALGORITHMS

Here's another problem: actual information theorists, who do understand what he's saying, think Dr. Dembski is nuts. Real information theorists often work with what they call "evolutionary algorithms" and "genetic algorithms" that mimic evolutionary processes. They do this as a means of finding excellent solutions to problems. They deliberately don't try to design a solution ahead of time, but let simulated evolutionary processes involving random mutations arrive at one.

Real information theorists wouldn't be doing what they're doing if someone in their own field had published a reliable paper showing that what they are doing is impossible.

So here's the interesting thing: don't you think that if Dembski, or anybody else, had proven, using information theory, that what his fellow information theorists were trying to do was impossible, that his fellow information theorists would have heard about it? Or at least noticed that their evolutionary algorithms didn't work? Unfortunately for Dr. Dembski, the folks who are the *real* practitioners of information theory haven't heard that

1. Dembski, *The Design Revolution*, 20.

Dr. Dembski has "proven" that what they routinely do is actually impossible. So they keep doing it, and it keeps working. Just like the bumblebees that kept flying.

So Dr. Dembski has managed to be wrong in two fields at once. Biology, in which he has done no experiments, and information theory, in which he has done no equations, formulas, theorems, or lemmas. In each case, he appears to have drawn conclusions based on no actual data at all. Being wrong in two fields at once is *inexcusable failure squared!*

As to the *real* human genome, that has its own problems . . . .

Chapter 33

Bad Design—the Human Genome

Baroque Design: Gratuitous Genome Complexities

"For natural theologians in centuries past, as well as for adherents to present-day versions of strict religious creationism, biotic complexity is the hallmark—the unquestionable signature—of ID. However, gratuitous or unnecessary biological complexity—as opposed to an economy of design—would seem to be the antithesis of thoughtful organic engineering. Yet, by objective scientific evidence, gratuitous and often-dysfunctional complexities (both in molecular structure and molecular operations) are so nearly ubiquitous as to warrant the status of hallmarks of the human genome."[1]

The human genome is a relatively newly understood area of *bad design*.

In his book *The Design Revolution*, Dr. Dembski claims that the human genome must have been designed, because it is too complex to have happened any other way. What a shame, then, that he didn't look at the real human genome. If he had, he would have seen something so messy and dangerous that it would never be attributed to a conscious designer, much less an intelligent one.

Here's why the real human genome is a case of *bad design*.

1. John C. Avise, "Footprints of nonsentient design inside the human genome," *Proceedings of the National Academy of Science*, vol, 107, suppl. 2, May 11, 2010, 8969–76, http://www.pnas.org/content/107/suppl.2/8969.full.pdf.

Try to imagine instructions for putting up a building that need to be copied many times, so that many different work crews can build their separate parts, all of which will have to fit together in the final building.

Then imagine that there are certain parts of the instructions that have to be followed exactly, in order to have a safe, usable building. For instance, the ventilation shafts have to fit together properly, the foundation has to be strong enough, the materials used in framing the building must be adequately strong, and the electrical wiring must be safe.

Then imagine that for every one copy of the electrical wiring diagram, you get five extra copies of blueprints for the ventilation shafts, not because you need them, but because that's the way the copies come out. This is the case with the human genome. This is *bad design*. The parts of our genome that are simply extra, unneeded copies of various genetic sequences are known as *duplicons*.

Imagine further that one piece of the directions for building and operating the furnace is in one part of the instructions, while the remaining pieces of those directions are located in a different part of the instructions. This is the case with the human genome. This is *bad design*. A piece of our genome is stored in parts of our cells called *mitochondria*. The rest of our genome is stored in the nuclei of our cells. In addition, the genetic structure in our mitochondria is different in various ways from the genetic structure in our cells' nuclei.

Imagine still further that some of these instructions are carried out first in one place, and then in another, with the project being stopped at a crucial point and transferred to the other location. What's more, in one location it is always windy and rainy, and the instructions frequently get damaged while they are there, but the building project has to go forward anyway, even though the instructions may have been damaged, and the building is therefore likely to be built incorrectly. Why would anybody make part of the building process take place in an unprotected area where crucial instructions get damaged as a matter of course? This is also the case with the human genome. This is *bad design*.

Our mitochondria have some, but not all, the genes that are needed for mitochondria to work. The remaining genes are in the cell's nucleus. Some crucial genes have to spend time being exposed to the cell's fluid-filled interior. This has a high concentration of chemicals that lead to mutations, that is, bad copying. Why would a wise designer do this?

Imagine further that certain parts of the instructions can copy themselves independently, and then insert those new copies into random places in the instructions without notifying anybody, and without any concern for what gets damaged or confused where it splices itself in.

Then imagine that sometimes, these independently-made copies insert themselves in such a way that some of the important instructions that they insert themselves into wind up getting damaged or even deleted as a consequence. This results in damage to or even failure of the resultant building. This is the case with our genome. This is *bad design*.

Imagine further that these rogue, independently copying elements made up 46 percent[2] of the total instructions for making the building.

This is the case with our genome. This is *bad design*.

These independently copying elements in our genome are known as *transposable elements*.

Then imagine further still that the process for copying the instructions for all the work crews is so bad in general, that mistakes in the copying happen on a regular basis, resulting in numerous buildings that are so bad that they are unsafe to use. Then imagine further that many of the fatal mistakes in copying could be avoided, if only a few simple changes to the copying processes had been made.

This is the case with the human genome. This is *bad design*.

The part of our genome that is like a set of building instructions is our DNA. Unfortunately, it is pretty easy for DNA to get copied incorrectly so that there are errors in the copy. This is called mutation.

MUTATIONS—A PART OF LIFE

Let's start with some basics. The human genome mutates. All the time. In our bodies, this can create cancer. In our reproductive cells, this can lead to sons and daughters with Huntington's chorea, ALS, hemophilia, and other awful and unnecessary genetic diseases. If human genetic material didn't mutate, then we wouldn't get all these diseases.

Mutations shouldn't exist in a designed system. They are far worse than useless in a system that has been designed. In manufacturing organizations around the world, systems are deliberately put into place to minimize

2. Victoria P. Belancio, Prescott L. Deininger, and Astrid M. Roy-Engel, "LINE dancing in the human genome: transposable elements and disease," *Genome Med.,* 2009, 1(10) 97, DOI: 10.1186/gm97.

random variation in products, and particularly to minimize dangerous variations. Human beings are pretty good at keeping random changes in their own work down to a bare minimum.

Yet somehow, this Designer who is powerful enough to have created us all hasn't figured out basic quality assurance. Photocopying machines work better than this!

EVOLUTION REQUIRES MUTATION

On the other hand, the basic mechanism of evolution is mutation, followed by natural selection. These lousy copying mechanisms make sense once you realize that *evolution could not take place without copying errors.*

That's how evolution works! That's how you get a new species, or even several new species, from one original species. Mutation is one of the fundamental drivers of evolution. It is, for better and worse, the reason that we exist as human beings. We got here because of errors in copying, followed by massive, relentless selection by nature. This selection determined which erroneous copies survived and copied themselves, and which erroneous copies died before they could reproduce.

So mutations are the hallmark of an evolved system. Without mutation, there would be no evolution. Dangerous mutations are an unfortunate part of evolution. Individuals who get dangerous mutations often die. Evolution, which is an unconscious process, doesn't care.

This is why mutations shouldn't exist in a designed system, but must in an evolved one. Without mutations, evolution wouldn't happen. But without mutations, a designed system would be *much* better. Our genome has numerous properties that encourage it to mutate frequently.

So what do you think? Did this mutation-friendly genome come about through evolution, which depends on mutation to work, or did it come about through an intelligent designer, who could have and should have done a lot more to make mutations rare and unusual?

MUTATIONS DRIVE EVOLUTION, BUT THEY ARE HARMFUL TO INDIVIDUALS

Although mutations are necessary for evolution, they can also be deadly. For instance, in every cell of your body, simple copying errors can cause cancer. In fact, the other word for something that causes cancer—a carcinogen—is a mutagen, that is, something that causes mutation. Almost half of all Americans develop cancer during their lifetimes, and a fifth will die from it.[3] All these are caused by incorrect copying of DNA. So if you know anyone who has had cancer, please remember that their disease was caused by our genome's tendency to get it's copying wrong. And copying itself is one of our genome's fundamental jobs.

What's more, when there are errors in the DNA copies that are located in sperm cells or egg cells, a baby resulting from them may have numerous genetic diseases, such as the hemophilia that would have killed Czar Nicholas II's son, if the Bolsheviks hadn't killed him first.

Transposable elements alone cause so much damage that they are responsible for a whole host of deadly human diseases, including hemophilia,[4] neurofibromatosis, choroideremia, cholinesterase deficiency, Apert syndrome, Dent's disease, β-thalassemia, Walker-Warburg syndrome, Duschenne muscular dystrophy, and retinitis pigmentosa.[5]

Mutations in general can debilitate the nervous system, liver, pancreas, bones, eyes, ears, skin, urinary and reproductive tracts, endocrine system, blood and other features of the circulatory system, muscles, joints, dentition, immune system, digestive tract, limbs, lungs, and almost any other body part you can name. Many of these problems will shorten a person's life, and about half will cause death before the age of thirty years. In many cases, the symptoms are horrific.

Why haven't the ID people heard of all these diseases?

Here's what happens to people with amyotrophic lateral sclerosis (ALS, also called Lou Gehrig's disease)—one of many diseases resulting from mutation. Ask yourself if this sounds like it was made by an intelligent designer: the nerves that tell your muscles what to do slowly die. As they die, the muscles they control get weaker and eventually die. As the disease progresses, the people lose control of their arms, legs, and body. Eventually, patients lose the ability to breathe. Along the way, they have trouble

---

3. Elaine N. Marieb, *Human Anatomy and Physiology,* 6th ed. (San Francisco: Pearson/Benjamin Cummings), 145.

4. B. A. Dombroski, S. L. Mathias, E. Nanthakumar, A. F. Scott, H. H. Kazazian Jr., "Isolation of an active human transposable element," *Science,* December 20, 1991, 254 (5039) 1805–8.

5. See Julia A. McMillan, Ralph D. Feigin, Catherine DeAngelis, and M. Douglas Jones, *Oski's Pediatrics: Principles & Practice,* 4th ed. (Philadelphia: Lippincott Williams and Wilkins, 2006).

speaking and swallowing, and their gag reflex becomes overactive. Most people die within three to five years of being diagnosed.

All because of one gene mutation.

One would think that God could do better.

Chapter 34

Bad Design—the Human Appendix

Now let's move on to one of our least desirable features—the human appendix.

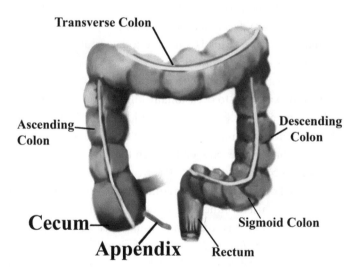

Figure 34.1   Human appendix and colon.

Here it is, the human appendix, or the vermiform appendix if you're a scientist.

It's an odd part of our digestive system, and it's particularly odd because it doesn't actually digest anything. It serves as a part of the digestive tract's immune system, but so does the rest of the digestive tract. Even *Gray's Anatomy* calls it a "functionless organ," and *Gray's Anatomy* is not known for its sense of humor. Having an organ that performs no particular function is *bad design.*

The appendix is a blind sack in which bacteria grow. In fact, it is a blind sack off of another blind sack called the cecum. A blind sack off of a blind sack is the perfect place for a bacterial colony.

In rabbits and other creatures that digest woody plants, the bacterial colonies that grow in the cecum and appendix can help the animal to digest wood. Humans don't digest wood. We have the organs for it, but they don't work. Having an organ in humans that works best in rabbits is *bad design.* Unfortunately, the appendix is still a great place for bacteria to breed. So every now and then, a colony of really nasty bacteria gets going in the human appendix. When this happens, the appendix gets a lethal infection called appendicitis and it must be surgically removed. If it is not removed, the person dies. Having an organ that performs no particular function and occasionally kills you is *really bad design.*

Each year, 500,000 people in the United States suffer from appendicitis. In the days before decent surgical techniques, people died regularly from infected appendixes. This is still true today in areas where modern surgical techniques cannot be practiced.

Since the appendix performs no critical function in humans, and its primary effect in existing is to occasionally kill us at any age for no reason, it is not clear why an infallible Creator would give us one. It is also hard to see why a Creator making us in his image would give us an organ that works best in rabbits. However, seen through the lens of evolution, the appendix is easier to comprehend.

It's a vestigial organ. That is, sometime in our vertebrate past, the appendix performed a useful function, and a functionless trace of that old, previously useful appendix is still with us today, *because it's not detrimental enough to kill us before childbirth, most of the time.* This is also the reason why we still have tails. We don't use them, but our ancestors did, and a little trace of a tail is still with us.

This is also why we get goose bumps when we are cold or frightened. We have tiny hairs that cover almost our entire body. There is a tiny muscle attached to each hair, and when we are cold or frightened, the muscle pulls the little hair into standing position. Why? Because in an animal with lots of fur, standing your fur up when you are cold increases the thickness of your coat and makes you warmer. When you are frightened, standing your fur on end makes you look bigger, which can help when facing down an enemy. We don't have lots of fur any more, so getting goose bumps from standing our hair on end doesn't make us any warmer, or make us look any bigger. But it reminds us of a time when our ancestors had real fur instead of tiny hairs, and this reaction, though now functionless, is still with us.

In case you don't believe me, I have pictures.

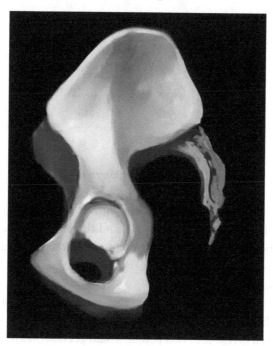

Figure 34.2   Adult human female pelvis.

Here is a human female pelvis, seen from the side. The left side is facing you. The tail is the last portion of the backbone, just as it is in all vertebrates. It is made up of bones called the sacrum and the coccyx, and it is present in all humans. It's a tail.

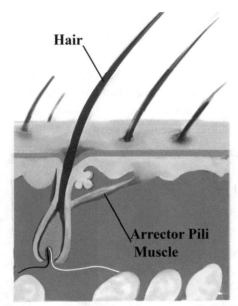

Hair

Arrector Pili Muscle

Figure 34.3   Human hair with *arrector pili* muscle.

Here is a picture of a human hair, with its tiny attached muscle. The muscle is called an *arrector pili* muscle, and there is one attached to every hair on our body. When these tiny muscles contract, you get goose bumps.

So this is why an appendix can exist in an evolved animal, but shouldn't in a well-designed one.

Now some people have claimed that the appendix would serve as a reservoir of beneficial gut bacteria for humans living in the natural environment, based on the idea that people would get diarrhea often enough to need a means of replacing their gut bacteria on a regular basis. This proposal does not hold water, however, since our guts are populated within hours of our birth. In addition, most modern human beings take courses of antibiotics on an occasional basis throughout their lives. Antibiotics kill most of the bacteria in the body, including the bacteria residing in the appendix, so it cannot serve as a bacterial reservoir during these times. Yet modern human beings, after a course of antibiotics, usually manage to repopulate their gastrointestinal tracts without undue problems, even without recourse to a seed population of bacteria coming from the appendix.

One textbook, however, offered an alternative opinion as to the true use of the vermiform appendix:

"Its major importance would
appear to be
financial support
of the surgical profession."

Figure 34.4   Alfred Sherwood Romer and Thomas S. Parsons, *The Vertebrate Body*,
6th ed. (Philadelphia: Saunders College Publishers, 1986), 389.

So perhaps God likes *surgeons* best.

Chapter 35

Evolution: The Greatest Indisputably True Story Ever Told

So there you have it. I have given you ten simple examples of how the human body is badly designed. I could have chosen hundreds. If I had looked at the rest of nature, I could have chosen thousands. I've pointed out a few of the myriad ways in which ID is a political pressure group that gets science wrong.

But now I want to talk about beauty.

I want to point out that the human body is actually wonderful. It's just that it's wonderful in the weird, crazy way that evolved systems are wonderful, rather than being wonderful in the careful, mathematical way that designed systems are. There's real beauty and utility here, but not pre-planned design.

I think that our bodies are beautiful the way they are, regardless of their imperfections.

But I also think that there is also real beauty in our ability to think, do research, and really understand the world we live in rather than just making up stories.

This capacity allows us to understand atoms and molecules that we can't even *see*, and principles we can't see that allow airplanes to fly, electrons we can't see that light up electric light bulbs, and yes, evolution, which often takes eons to occur, so no one human being can see it working, but we understand it and *we can make successful predictions based on it*. So we know it's there.

ID does not make predictions. Evolutionary theory does. The evolution of antibiotic-resistant bacteria can only be understood by evolutionary theory. Antibiotic-resistant bacteria are now killing people we used to be able to cure.

I fear having public health officials who don't believe in evolution.

Ecology—which is the interplay between different plants and animals and their environments—can only be understood if you understand evolution.

I fear having environmental policy made by people who don't believe in evolution.

I realize that many people don't like the fact that there are unanswered questions in evolution. However, there are unanswered questions in every scientific field. Modern physicists don't know whether gravity is smooth and continuous, or comes in little particles. Yet we use modern physics every day to launch satellites, and they stay in orbit. Biblical physics couldn't do that.

You don't throw out an entire scientific field because there are a few unanswered questions. In fact, the whole point behind science is to investigate unanswered questions. That's what research is all about. If there are no unanswered questions then there is no research. If there are no unanswered questions, then there is no hope for improvement. If there are no unanswered questions, then the world is a far duller place. Unanswered questions are why scientists go to work every morning. Awe and wonder are as much a part of science as they are a part of any religion.

I realize that many people dislike what they think are the implications of evolution. "If it's all about eating and breeding and dying," they wonder, "where do beauty and meaning and justice fit in? Where does purpose fit in?"

Where indeed? One answer is that they fit in where we decide to put them. If we human beings think that beauty and meaning and justice and purpose are important, then it is up to us to put them into our lives and into our world, rather than waiting for God to hand them to us on a silver platter.

I realize that other people may be afraid to accept evolution because they want to believe in the uniqueness of human beings. Fortunately, science tells us that we as a species are indeed unique. So is every other species. What's more, it's clear that we are not particularly favored by any god or gods, whatever our egos may tell us to the contrary. Does that mean that we're worthless? Far from it. We can experience awe and wonder. We have an amazing ability to figure things out. We can make art and appreciate beauty. We can build cities yet appreciate wilderness. We can strive to be the best that we can be and to build the best society that we can build. We can require justice, in the firm knowledge that no one has any divine rights. And we can love each other and love this planet.

I realize that in writing this book, I may be preaching to the choir. However, our public discourse on evolution needs simple, straightforward arguments in order to counter the false claims made by proponents of ID. In other words, we need talking points. Ones that we can bring to our legislators, school boards, and politicians of all stripes. I have tried to provide talking points that anybody can use right here in this book.

So there are times when preaching to the choir is exactly the right thing to do. We need clarification, reassurance, and solid political-style arguments if we are going to defend evolution against pseudoscientists and political pressure groups. In fact, my main hope for this book is that I've given you, the choir, some great new songs to sing. And in all cases, keep in mind that evolution is the greatest *indisputably true* story ever told.

Figure 35.1   Our place in the galaxy.

Phylogenetic Tree of Life

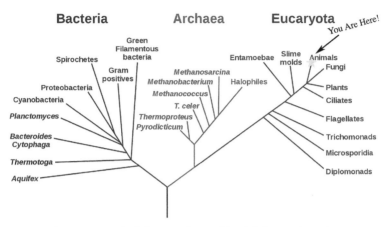

Figure 35.2   A phylogeny of everything (phylogenetic tree).
Or, our place on the family tree.

Appendix 1

The Phylogeny of Mammalian Testicles

External testicles are owned only by the boreoeutherian land mammals. This large group of mammals includes humans. These testes produce sperm best at temperatures that are lower than core body temperature. For this reason, they are located outside the body in the scrotum. However, there are exceptions. For instance, rhinoceroses, tapirs, bats, and marine mammals such as seals, dolphins, and manatees are all boreoeutherian mammals, but their testes do not hang outside the body.

Non-boreoeutherian mammals don't have external testicles, either. These include monotremes (echidnas and platypuses), armadillos, sloths, and elephants.

Here's a chart.

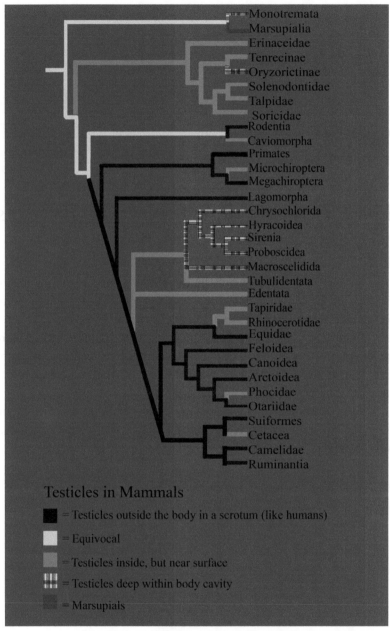

Figure A1.1   Phylogeny of mammalian testicles.
The evolution of the scrotum and testicular descent in mammals: a phylogenetic view.
Source: Werdelin, L, Nilsonne, A. Journal of Theoretical Biology 1999 Jan 7;196(1): 61–72.
Department of Palaeozoology, Swedish Museum of Natural History, Box 50007, S-104 05
Stockholm, Sweden. werdelin@nrm.se

ABSTRACT

The adaptive significance of the scrotum and the evolution of the descent of the testicles and epididymis have been a focus of interest among biologists for a long time. In this paper we use three anatomical character states of the scrotum and descensus: (1) testicles descended and scrotal; (2) testicles descended but ascrotal; (3) testicles not descended (testicondy). These states are then mapped on an up-to-date phylogeny of the Mammalia. Three main points arise out of this mapping procedure: (1) the presence of a scrotum is either primitive in extant Mammalia or primitive within eutherian mammals except Insectivora; (2) evolution has generally proceeded from a scrotal condition to progressively more ascrotal; (3) loss of testicular descensus is less common in mammalian evolution than is loss of the scrotum. In the light of these findings we discuss some current hypotheses regarding the origin and evolution of the scrotum. We find that these are all incomplete in so far as it is not the presence of the scrotum in various mammal groups that requires explaining. Instead, it is the reverse process, why the scrotum has been lost in so many groups, that should be explained. We suggest that the scrotum may have evolved before the origin of mammals, in concert with the evolution of endothermy in the mammalian lineage, and that the scrotum has been lost in many groups because descensus in many respects is a costly process that will be lost in mammal lineages as soon as an alternative solution to the problem of the temperature sensitivity of spermatogenesis is available.

PMID: 9892556 [PubMed—indexed for MEDLINE]

Appendix 2

The American Association for the Advancement of Science's Statement on Intelligent Design

AAAS BOARD RESOLUTION ON INTELLIGENT DESIGN THEORY

The contemporary theory of biological evolution is one of the most robust products of scientific inquiry. It is the foundation for research in many areas of biology as well as an essential element of science education. To become informed and responsible citizens in our contemporary technological world, students need to study the theories and empirical evidence central to current scientific understanding.

Over the past several years proponents of so-called "intelligent design theory," also known as ID, have challenged the accepted scientific theory of biological evolution. As part of this effort they have sought to introduce the teaching of "intelligent design theory" into the science curricula of the public schools. The movement presents "intelligent design theory" to the public as a theoretical innovation, supported by scientific evidence, that offers a more adequate explanation for the origin of the diversity of living organisms than the current scientifically accepted theory of evolution. In response to this effort, individual scientists and philosophers of science have provided substantive critiques of "intelligent design," demonstrating significant conceptual flaws in its formulation, a lack of credible scientific evidence, and misrepresentations of scientific facts.

Recognizing that the "intelligent design theory" represents a challenge to the quality of science education, the Board of Directors of the AAAS unanimously adopts the following resolution:

Whereas, ID proponents claim that contemporary evolutionary theory is incapable of explaining the origin of the diversity of living organisms;

Whereas, to date, the ID movement has failed to offer credible scientific evidence to support their claim that ID undermines the current scientifically accepted theory of evolution;

Whereas, the ID movement has not proposed a scientific means of testing its claims;

Therefore Be It Resolved, that the lack of scientific warrant for so-called "intelligent design theory" makes it improper to include as a part of science education;

Therefore Be Further It Resolved, that AAAS urges citizens across the nation to oppose the establishment of policies that would permit the teaching of "intelligent design theory" as a part of the science curricula of the public schools;

Therefore Be It Further Resolved, that AAAS calls upon its members to assist those engaged in overseeing science education policy to understand the nature of science, the content of contemporary evolutionary theory and the inappropriateness of "intelligent design theory" as subject matter for science education;

Therefore Be Further It Resolved, that AAAS encourages its affiliated societies to endorse this resolution and to communicate their support to appropriate parties at the federal, state and local levels of the government.

Approved by the AAAS Board of Directors on 10/18/02

Appendix 3

The Life Cycle of the Immortal Jellyfish

Turritopsis nutricula is the first case in which a metazoan is capable of reverting completely to a sexually immature, colonial stage after having reached sexual maturity as a solitary stage. Thus, it appears that it has cheated death and is a potentially immortal, solitary metazoan.

The Life Cycle of the Immortal Jellyfish

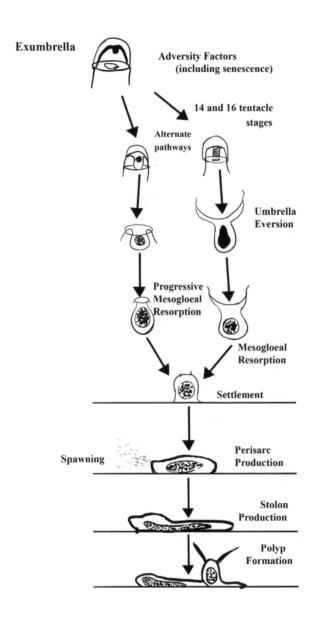

Figure A3.1   The repeating life cycle of the immortal jellyfish (*Turritopsis nutricula*).

Appendix 4

Credits and Notes for Illustrations and Photographs

CHAPTER 2: BAD DESIGN—MEN'S TESTICLES

Illustration: Reproductive system of the male human.
Original drawing by: Alexander Winkler
Copyright: Abby Hafer

Illustration: Reproductive system of the male frog.
Original drawing by: Alexander Winkler
Copyright: Abby Hafer

CHAPTER 4: TESTICLES, PART 2

Illustration: Elephants have their testicles deep within their bodies.
Original drawing by: Alexander Winkler
Copyright: Abby Hafer

Illustration: Birds do too.
Original drawing by: Alexander Winkler
Copyright: Abby Hafer

CHAPTER 6: INTELLIGENT DESIGN ACCORDING TO ITS BELIEVERS: IS INTELLIGENT DESIGN THE SAME AS CREATIONISM, AND IS INTELLIGENT DESIGN RELIGION?

Copies of the two texts:

Image caption: *Of Pandas and People* (1987, creationist version), ch. 3, p. 40.

Image caption: *Of Pandas and People* (1987, "intelligent design" version), ch. 3, p. 41.

> Two versions of the book *Of Pandas and People* were used in the lawsuit referred to as *Kitzmiller v. Dover Area School District*. This was the seminal case regarding teaching ID in public school science classes in Dover, Pennsylvania. These two particular excerpts from those books were not used during the trial, but were made available to the public after the trial. They can be found at the website for the National Center for Science Education, http://ncse.com/creationism/legal/cdesign-proponentsists.

CHAPTER 7: THE INFAMOUS WEDGE STRATEGY

Image: Copy of the cover of *The Wedge Strategy*.
The entire Wedge Strategy document, including the cover picture, can be found at http://ncse.com/files/pub/creationism/The_Wedge_Strategy.pdf.

CHAPTER 8: WHY DENYING EVOLUTION CAN GET YOU INTO TROUBLE AND CAUSE MASS HUNGER, TOO

Photograph: Lysenko with Khrushchev.
Photograph caption: Lysenko (far left) with Nikita Khrushchev (center left), the Soviet leader. Clearly, Lysenko was favored by Communist Party leadership.
Photograph of Nikita Khrushchev, Trofim Lysenko, and others
Original photograph by: Unknown
Gray scaling and cropping by Gretjen Hargesheimer

CHAPTER 11: BAD DESIGN—THE BIRTH CANAL

Illustration: A baby crowning.
Original drawing by: Alexander Winkler
Copyright: Abby Hafer
After the Boston Women's Health Book Collective, *Our Bodies Ourselves—A Book By and For Women* (New York: Simon and Schuster, 1973).

Illustration: The birth process.
Original drawing by: Alexander Winkler
Copyright: Abby Hafer
After the Boston Women's Health Book Collective, *Our Bodies Ourselves—A Book By and For Women* (New York: Simon and Schuster, 1973).

Photograph caption: Kangaroo and joey.
Original title/description of photograph: An Eastern Grey Kangaroo with a joey in her pouch.
Photo taken November, 2005
Namadgi National Park, Australia
Photograph credit: Photo taken by Martyman
Copyright information: Creative Commons Attribution-ShareAlike 3.0 License.
Gray scaling and cropping by Gretjen Hargesheimer

Illustration: A healthy female human reproductive system.
Illustration caption:
Original drawing by: Alexander Winkler
Copyright: Abby Hafer

Illustration: A female human reproductive system with a fistula.
Illustration caption: An obstetric fistula is a passage that forms between the birth canal and either the rectum or the urinary bladder. It is the result of a difficult labor in which some of the vaginal tissue is killed in the process of giving birth. This illustration shows a fistula between the rectum and the birth canal.
Original drawing by: Alexander Winkler

Copyright: Abby Hafer

CHAPTER 12: THE HANDY-DANDY INTELLIGENT DESIGN REFUTER, PART 1

Illustration: Text of the original sticker made by the Cobb County, Georgia Board of Education.
Original drawing by: Alexander Winkler
Copyright (but not of the original sticker): Abby Hafer

Illustration: An imagined "spherical-earth controversy" sticker.
Illustration caption: Reproduction of the first sticker with new text.
Original drawing by: Alexander Winkler
Text provided by Abby Hafer
Copyright: Abby Hafer

CHAPTER 14: ANIMALS THAT SHOULDN'T EXIST ACCORDING TO INTELLIGENT DESIGN

Photograph: Mudskipper climbing onto a rock.
Photographer's title: Mudskipper (*Periophthalmodon septemradiatus*) in a tank, climbing onto a rock
Photograph courtesy of Gianluca Polgar
Copyright: Gianluca Polgar
Gray scaling and cropping by Gretjen Hargesheimer

Photograph: A mudskipper taking a walk on the beach.
Photographer's title: Mudskipper (*Boleophthalmus boddarti*) taking a walk on the mud bank of a tidal creek
Photograph courtesy of Gianluca Polgar
Copyright: Gianluca Polgar
Gray scaling and cropping by Gretjen Hargesheimer

APPENDIX 4

Photograph: A mudskipper climbing a tree.

Photographer's title: Mudskipper (*Periophthalmus chrysospilos*) perching on a root of a mangrove tree

Photograph courtesy of Gianluca Polgar
Copyright: Gianluca Polgar

Gray scaling and cropping by Gretjen Hargesheimer

Photograph: A mudskipper's special fused pelvic fin that helps with climbing.

Photographer's title: Mudskipper's special fused pelvic fins (*Periophthalmus chrysospilos*) that help with climbing

Photograph courtesy of Gianluca Polgar

Copyright: Gianluca Polgar

Gray scaling and cropping by Gretjen Hargesheimer

Photograph: A mudskipper's nesting burrow.

Photograph caption: You can see the two eyes peeking out of the hole in the sand.

Photographer's title: Mudskipper's reproductive burrow (*Periophthalmus malaccensis*)

Photograph courtesy of Gianluca Polgar

Copyright: Gianluca Polgar

Gray scaling and cropping by Gretjen Hargesheimer

Photograph: *Turritopsis nutricula,* the immortal jellyfish.

Original title/description: *Turritopsis nutricula*

Attribution: Alvaro Esteves Migotto

Gray scaling and cropping by Gretjen Hargesheimer

Copyright information: Permission to use this photograph in this book was given to Abby Hafer by the photographer, Alvaro E. Migotto.

> Note on photograph: "When the reversal of life cycle was published *T. nutricula* was considered a cosmopolitan species, but recent data on morphology and DNA have shown that several of the nominal *Turritopsis* species are valid. The species of those papers were in fact *Turritopsis dohrnii*, as they were collected from the Mediterranean, where it is found. *Turritopsis nutricula* is found in the Caribbean and the Americas. As far as I know the life cycle reversal was not yet observed in *T. nutricula*! But I guess it will soon…"

Alvaro Esteves Migotto.

Centro de Biologia Marinha, Universidade de São Paulo, SP, Brasil

CHAPTER 15: BAD DESIGN—THE HUMAN THROAT

Illustration: Human esophagus and trachea.

Illustration caption:The pathway for air is marked with dots. The pathway for food and liquids is marked with lines. *Note:* these paths overlap in the mouth, and cross at the pharynx.

Original drawing by: Alexander Winkler

Copyright: Abby Hafer

Illustration: Tracheal blockage due to inhalation of food.

Illustration caption: This results in an inability to breathe.

Original drawing by: Alexander Winkler

Copyright: Abby Hafer

Photograph: A whale spouting.

Original title: North Atlantic Right Whale mother and calf *(Eubalaena glacialis)*

Attribution: Florida Fish and Wildlife Conservation Commission/NOAA

Copyright information: This work is in the public domain in the United States because it is a work prepared by an officer or employee of the United States Government as part of that person's official duties under the terms of Title 17, Chapter 1, Section 105 of the US Code.

Gray scaling and cropping by Gretjen Hargesheimer

CHAPTER 17: IRREDUCIBLE COMPLEXITY, THE DESIGN INFERENCE, AND GEOLOGICAL FORMATIONS

Photograph: This natural arch is in the Sahara Desert.

Photograph caption: It is over six feet high, and its thinnest leg is only six inches thick. If you remove even a small part of the thinnest leg, it will collapse, so it is irreducibly complex. It was formed by wind erosion.

Attribution: Photo by Guilain Debossens

Copyright: Guilain Debossens

Gray scaling and cropping by Gretjen Hargesheimer

Photograph: Another natural arch.

Photograph caption: Look, a bridge! This must have been designed, right? Wrong. This bridge spans the Ardèche river in France. Although it is nearly 200 feet wide and nearly 150 feet high, and has no visible support from below, it is a natural formation.

Original title/description: The Pont D'Arc

Attribution: Paste

Copyright: This work is licensed under the Creative Commons Attribution-ShareAlike 3.0 License

Gray scaling and cropping by Gretjen Hargesheimer

Composite of four photographs

Composite photograph: Streamlined shapes.

Composite photograph caption: From top to bottom: penguin, shark, torpedo, whale.

Composite photograph of penguin, torpedo, shark and whales done by

Gretjen Hargesheimer

The individual photographs that make up this composite are credited as follows:

Penguin:

Original title: *Pygoscelis papua* —Gentoo Penguin swimming underwater at Nagasaki Penguin Aquarium, Nagasaki, Japan.

Attribution: Ken Funakoshi

Copyright information: This work is licensed under the Creative Commons Attribution-Share Alike 2.0 Generic license

Gray scaling and cropping by Gretjen Hargesheimer

Torpedo:

Credits and Notes for Illustrations and Photographs

Original title/description: Torpedo mark 37e of Israel.

Attribution: Natan Flayer

Copyright information: This file is licensed under the Creative Commons Attribution-Share Alike 3.0 Unported license.

Gray scaling and cropping by Gretjen Hargesheimer

Shark:

Original title/description: Sandbar shark (*Carcharhinus plumbeus*) at the Georgia Aquarium

Attribution: Brian Gratwicke

Copyright information: This file is licensed under the Creative Commons Attribution 2.0 Generic license.

Gray scaling and cropping by Gretjen Hargesheimer

Whales:

Original title/description: A mother sperm whale and her calf off the coast of Mauritius.

Attribution: Gabriel Barathieu

Copyright information: This work is licensed under the Creative Commons Attribution-Share Alike 2.0 Generic license

Gray scaling and cropping by Gretjen Hargesheimer

Photograph: A unique formation.

Photograph caption: This rock formation is utterly unique. It was not designed.

Original title: Balanced Rock

Attribution: Clarence Bisbee

Full photo credit: Twin Falls Public Library, Twin Falls, Idaho

Photograph Number 85

Twin Falls Public Library Photo Collection

Photographer: Clarence Bisbee

Copyright information: Copyright for this photograph is owned by the Twin Falls Public Library, Twin Falls, Idaho

Cropping by Gretjen Hargesheimer

Photograph: A symmetrical formation.

Photograph caption: Symmetry and a seemingly human form do not make this formation in Utah the work of a Designer.

Original title/description: Toadstool shaped hoodoo at Grand-Staircase Escalante National Monument rimrocks, Utah.

Attribution: Ciar

Copyright information: This file is licensed under the Creative Commons Attribution-Share Alike 3.0 Unported license.

Gray scaling and cropping by Gretjen Hargesheimer

Photograph: A round-sided tunnel in solid stone.

Photograph caption: Look! A smooth-sided tunnel through solid rock. This formation is found at Volcano National Park in Hawaii. It was not designed.

Original title/description: Hawaii: Thurston Lava Tube in the Volcano National Park

Attribution: Eli Duke

Copyright information: Creative Commons Attribution-ShareAlike 2.0 Generic (CC BY-SA 2.0)

Gray scaling and cropping by Gretjen Hargesheimer

CHAPTER 18: IRREDUCIBLE COMPLEXITY AND BLOOD CLOTTING

Photograph: Pineapple sea cucumber.

Photograph caption: In sea cucumbers, the animal constricts muscles near the injured area in order to reduce blood loss.

Original title/description: *Thelenota ananas*, more commonly known as the pineapple sea cucumber.

Attribution: Dan Norton

Copyright information: Copyright 2010 Dan Norton

Gray scaling and cropping by Gretjen Hargesheimer

Photograph: Sand dollars.

Photograph caption: In sand dollars, a temporary plug is formed when blood cells clump together.

Credits and Notes for Illustrations and Photographs

Original title/description: A group of Sand dollars in Monterey Bay Aquarium.

Attribution: Chan siuman

Copyright information: This file is licensed under the Creative Commons Attribution-Share Alike 3.0 Unported license

Gray scaling and cropping by Gretjen Hargesheimer

Composite of three photographs

Composit photograph: Sea stars.

Composite photograph caption: In sea stars, the clumped blood cells fuse together to form a clot, and are no longer individual cells.

Composite of three sea star photographs done by Gretjen Hargesheimer done by Gretjen Hargesheimer

1) Original title/description: Northern Pacific sea star, *Asterias amurensis*, dorsal view.

Attribution: Unknown

Copyright information: Public domain

Gray scaling and cropping by Gretjen Hargesheimer

2) Original title/description: Crown of Thorns (*Acanthaster planci*). Koh Similan, Boulder City, Thailand

Attribution: Jon Hanson

Copyright information: This file is licensed under the Creative Commons Attribution-Share Alike 2.0 Generic license.

Gray scaling and cropping by Gretjen Hargesheimer

3) Original title/description: *Protoreaster nodosus*, Oreasteridae—Kambodscha/Cambodia, Insel Kaoh Ruessei SSE Sihanoukville

Attribution: Franz Xaver

Copyright information: This file is licensed under the Creative Commons Attribution-Share Alike 3.0 Unported license

Gray scaling and cropping by Gretjen Hargesheimer

Photograph: Horseshoe crab.

Photograph caption: In horseshoe crabs, the blood cells form a clot and also spread out fibers. These fibers increase the size and scope of the clot and allow it to trap invading cells.

Original title/description: *Limulus polyphemus*—Atlantic Horseshoe Crab, American Horseshoe Crab

Attribution: Robert Pos

Source Collection: US Fish and Wildlife Service Online Digital Media Library

Copyright information: This media file is in the public domain.

Gray scaling and cropping by Gretjen Hargesheimer

Photograph: Vertebrates.

Photograph caption: Vertebrates like zebras and humans have blood that both forms clots and makes fibers. But the chemicals that make the fibers in our blood clots are completely different from the chemicals that do this in horseshoe crabs.

Original title/description: Plains Zebras (*Equus quagga*), more specifically the Damara subspecies (*Equus quagga antiquorum*) in Okavango, Botswana in 2002.

Attribution: Taken and submitted by Paul Maritz (Paulmaz)

Copyright information: This file is licensed under the Creative Commons Attribution-Share Alike 3.0 Unported license

Gray scaling and cropping by Gretjen Hargesheimer

CHAPTER 19: BAD DESIGN—THE HUMAN BLOOD CLOTTING SYSTEM: IT LED TO THE COMMUNIST RUSSIAN REVOLUTION.

Photograph: Alexei.

Original title: Alexei Nikolaevich—Tsarevich of Russia

Attribution: Photograph credit: Bain News Service, publisher

Copyright information: This media file is in the public domain in the United States. This applies to US works where the copyright has expired, often because its first publication occurred prior to January 1, 1923.

Cropping by Gretjen Hargesheimer

Photograph: Rasputin.

Original title: Rasputin

Attribution: unknown

Copyright information: This work is in the public domain in Russia according to article 1256 of the Civil Code of the Russian Federation.

This work was published on territory of the Russian Empire (Russian Republic) except for territories of the Grand Duchy of Finland (*Великое княжество Финляндское*) and Congress Poland (*Царство Польское*) before 7 November 1917 and wasn't republished for 30 days following initial publications on the territory of Soviet Russia or any other states.

The Russian Federation (early RSFSR, Soviet Russia) is the historical heir but not legal successor of the Russian Empire.

Cropping by Gretjen Hargesheimer

CHAPTER 21: IRREDUCIBLE COMPLEXITY AND THE HUMAN EYE

Illustration: The human eye.
Original drawing by: Alexander Winkler
Copyright: Abby Hafer

Illustration: Diagram of the human retina.
Illustration caption: This diagram shows the pathway that light must take to reach the photoreceptors. Note the nerve fibers and multiple layers of cells through which the light must pass before making contact with the photoreceptors.
Original drawing by: Alexander Winkler
Copyright: Abby Hafer

Illustration: The closed-lens eye of *Pecten maximus.*
Original drawing by: Alexander Winkler
Copyright: Abby Hafer
After Luitfried von Salvini-Plawen and Ernst Mayr, "On the evolution of photoreceptors and eyes," *Evolutionary Biology,* vol. 10, edited by Max K. Hecht, William Steere, and Bruce Wallace (New York: Plenum, 1977), 207–63

Illustration: The closed-lens eye of *Cardium muticum.*

Original drawing by: Alexander Winkler

Copyright: Abby Hafer

After Luitfried von Salvini-Plawen and Ernst Mayr, "On the evolution of pho-toreceptors and eyes," in *Evolutionary Biology* vol. 10, edited by Max K. Hecht, William Steere, and Bruce Wallace (New York: Plenum, 1977), 207–63

Illustration: The closed-lens eye of *Tridacna maxima.*

Original drawing by: Alexander Winkler

Copyright: Abby Hafer

After Luitfried von Salvini-Plawen and Ernst Mayr, "On the evolution of pho-toreceptors and eyes," in *Evolutionary Biology,* vol. 10, edited by Max K. Hecht, William Steere, and Bruce Wallace (New York: Plenum, 1977), 207–63

Illustration: The pinhole eye of *Lima squamosa.*

Original drawing by: Alexander Winkler

Copyright: Abby Hafer

After Luitfried von Salvini-Plawen and Ernst Mayr, "On the evolution of pho-toreceptors and eyes," in *Evolutionary Biology,* vol. 10, edited by Max K. Hecht, William Steere, and Bruce Wallace (New York: Plenum, 1977), 207–63.

Illustration: The eye pit of *Arca barbata.*

Original drawing by: Alexander Winkler

Copyright: Abby Hafer

After Luitfried von Salvini-Plawen and Ernst Mayr, "On the evolution of pho-toreceptors and eyes," in *Evolutionary Biology,* vol. 10, edited by Max K. Hecht, William Steere, and Bruce Wallace (New York: Plenum, 1977), 207–63.

Illustration: Gastropod eyes.

Original drawing by: Alexander Winkler

Copyright: Abby Hafer

After Luitfried von Salvini-Plawen and Ernst Mayr, "On the evolution of pho-toreceptors and eyes," in *Evolutionary Biology,* vol. 10, edited by Max K. Hecht, William Steere, and Bruce Wallace (New York: Plenum, 1977), 207–63.

Illustration: Unicellular *Euglena* with eye spot and chloroplasts.

Illustration caption: The eyespot, a cluster of pigment, helps the organism sense and respond to light. The organism then moves toward light using its flagellum for locomotion. This way, the chloroplasts are properly positioned for photosynthesis. This allows the cell to make food.

Original drawing by: Alexander Winkler

Copyright: Abby Hafer

Illustration: Half an eye and a simple eye

Original drawing by: Alexander Winkler

Copyright: Abby Hafer

CHAPTER 22: BAD DESIGN—THE HUMAN EYE

Illustration: Find your blind spot.

Exercise made by Abby Hafer

Photograph: Photograph of the human retina, taken with an ophthalmoscope.

Photograph caption: Notice the blood vessels lying on top of the retina, which block the path of the light that's traveling to the photoreceptors.

Original title/description: Fundus photograph of the left eye, showing a fundus with no sign of disease or pathology.

Attribution: Mikael Häggström

Copyright information: This file is made available under the Creative Commons CC0 1.0 Universal Public Domain Dedication

Gray scaling and cropping by Gretjen Hargesheimer

Illustration: Diagram of the human retina and optic nerve.

Illustration caption: Notice the blood vessels, nerve fibers, and many layers of cells that are blocking the path of the light to the photoreceptors. The nerve fibers and blood vessels on top of the retina converge before entering the optic nerve, blocking the light entirely. This forms the human eye's blind spot.

Original drawing by: Alexander Winkler

Copyright: Abby Hafer

Photograph: Cuttlefish.

Original title/description: Cuttlefish

Attribution: David Sim

Copyright information: This file is licensed under the Creative Commons Attribution 2.0 Generic license.

Gray scaling and cropping by Gretjen Hargesheimer

Illustration: Diagrams of human and cephalopod eyes.

Original drawing by: Alexander Winkler

Copyright: Abby Hafer

CHAPTER 25: BAD DESIGN—OUR BIOCHEMICAL PATHWAYS. IF CATS DON'T DIE OF SCURVY, THEN WHY DO WE?

Photograph: Why is this cat smiling?

Photograph caption: Because it can't get scurvy

Original title/description: Laughing cat (LOLCAT.JPG)

Attribution: Koruko

Copyright information: This file is licensed under the Creative Commons Attribution-Share Alike 3.0 Unported license.

Gray scaling and cropping by Gretjen Hargesheimer

Photograph: Symptoms of scurvy: gums.

Photograph caption: Gums redden, recede and blacken. Teeth loosen and fall out.

Original title/description: This patient presented with scorbutic gums due to a vitamin C deficiency, a symptom of scurvy.

Attribution: Centers for Disease Control, Public Health Image Library Image #3998

Copyright information: This image is in the public domain and thus free of any copyright restrictions. As a matter of courtesy we request that the content provider be credited and notified in any public or private usage of this image.

Gray scaling and cropping by Gretjen Hargesheimer

Credits and Notes for Illustrations and Photographs

Photograph: Scorbutic tongue.

Photograph caption: Small blood vessels just under the surface of the tongue break and release blood.

Original title/description: This patient presented with a "scorbutic tongue" due to what proved to be a vitamin C deficiency.

Attribution: Centers For Disease Control, Public Health Image Library Image #6239

Copyright information: This image is in the public domain and thus free of any copyright restrictions. As a matter of courtesy we request that the content provider be credited and notified in any public or private usage of this image.

Gray scaling and cropping by Gretjen Hargesheimer

Photograph: Symptoms of scurvy: skin.

Photograph caption: Blood vessels break and bleed under the skin.

Original title/description: This patient presented with a case of perifollicular petechiae of the skin caused by a vitamin C and/or vitamin K deficiency.

Attribution: Centers For Disease Control, Public Health Image Library Image #6238

Copyright information: This image is in the public domain and thus free of any copyright restrictions. As a matter of courtesy we request that the content provider be credited and notified in any public or private usage of this image.

Gray scaling and cropping by Gretjen Hargesheimer

Photograph: Symptoms of scurvy: legs, back.

Photograph caption: We get rough skin and blotchy bruises all over the body, particularly on the legs. Bruises and exertion can result in internal bleeding. Legs and arms can swell up.

Original title/description: Back View of a Male Scurvy Victim (photograph was cropped in this iteration).

Attribution: National Library of Medicine

In: Skorbut in veroeffentlichungen, 1920, Aschoff & Koch.

Copyright information: The National Library of Medicine believes this item to be in the public domain.

Cropping by Gretjen Hargesheimer

Photograph: This cat didn't eat fresh produce. It didn't get scurvy.

Original title/description: Laughing cat on the roof of an apartment in Daratsos, Kriti, Greece (original title: Laughing Cat)

Attribution: DrPete (Pete Coleman)

Copyright information: This file is licensed under the Creative Commons Attribution 2.0 Generic license.

Gray scaling and cropping by Gretjen Hargesheimer

Photograph: This man didn't eat fresh produce. He did get scurvy.

Original title/description: Back View of a Male Scurvy Victim

Attribution: National Library of Medicine

In: Skorbut in veroeffentlichungen, 1920, Aschoff & Koch.

Copyright information: The National Library of Medicine believes this item to be in the public domain.

Cropping by Gretjen Hargesheimer

Photograph: Rat.

Original title/description: Pet rat named Albertina

Attribution: Dawn Huczek

Copyright information: This file is licensed under the Creative Commons Attribution 2.0 Generic license.

Gray scaling and cropping by Gretjen Hargesheimer

CHAPTER 26: WHY DOES INTELLIGENT DESIGN ACT SO MUCH LIKE THE TOBACCO LOBBY?

Screenshot: Event announcement from the Discovery Institute.

Screenshot caption: Here is the announcement for the event denying human-caused global warning. The event took place at the Discovery Institute. The coauthor of the book being presented is Fred Singer.

Title of Event: *Unstoppable Global Warming: Every 1,500 Years*

Date of Event: December 5, 2006

CHAPTER 27: BAD DESIGN—OUR TEETH, OR, WHY IS THIS ANIMAL SMILING?

Photograph: Q: Why is this shark smiling?

Photograph caption A: Because it doesn't have to go to the dentist.

Original title/description: Frilled shark (*Chlamydoselachus anguineus*)

Attribution: OpenCage (http://opencage.info/pics.e/large_13408.asp)

Copyright information: This file is licensed under the Creative Commons Attribution-Share Alike 2.5 Generic license.

Gray scaling and cropping by Gretjen Hargesheimer

Illustration: A diagram of a healthy tooth.

Original drawing by: Alexander Winkler

Copyright: Abby Hafer

Illustration: A diagram of a tooth with decay and infected pulp.

Illustration caption: Here is a diagram of a tooth in which bacteria have gotten into the soft inner core of the tooth. Notice how this infected core leads to your blood vessels—which lead to the rest of your body! No wonder people with bad teeth die young.

Original drawing by: Alexander Winkler

Copyright: Abby Hafer

Photograph: Why does this animal deserve a tooth pain-free life?

Original title/description: Shark

Attribution: Jeff Kubina

Copyright information: Creative Commons Attribution-ShareAlike 2.0 Generic (CC BY-SA 2.0)

Gray scaling and cropping by Gretjen Hargesheimer

Photograph: "This won't hurt a bit."

Original title/description: Great White Shark. Slash Attacks! New Zealand

Attribution: Lwp Kommunikacio

Copyright information: Creative Commons Attribution 2.0 Generic license. (CC by 2.0)

Gray scaling and cropping by Gretjen Hargesheimer

Photograph: The size of a *megalodon* tooth is frightening.

Original title/description: Dent fossilisée d'un requin *Carcharodon megalodon*, provenant du désert d'Atacama, au Chili, et datant du Miocène. Dimensions : 18 cm de diagonale et 13 cm de base.

Attribution: Lonfat

Copyright information: Public domain.

Gray scaling and cropping by Gretjen Hargesheimer

Photograph: So is the size of a *megalodon* jaw.

Original title/description: Jaws of a Megalodon, an ancient giant shark, on display at the National Aquarium, Napier, New Zealand. The jaws are not original fossils, but some kind of cast/mock-up. For scale, the man in the picture is 5'10"/1.78m.

Attribution: Original uploader was Jasper33 at en.wikipedia

Copyright: This work has been released into the public domain by its author, Jasper33 at the wikipedia project. This applies worldwide.

Gray scaling and cropping by Gretjen Hargesheimer

CHAPTER 28: THE DISCOVERY INSTITUTE HASN'T DISCOVERED ANYTHING (SORT OF LIKE THE TOBACCO RESEARCH INSTITUTE)

Chart: Histogram of word frequencies.

Chart caption: The frequency of the use of the words *argu-* and *data* by proponents of intelligent design at the Discovery Institute, and by scientists at the Smithsonian Tropical Research Institute (STRI).Scientists at STRI made heavy use of data, but rarely used or cited argumentation in their articles. Discovery Institute writers rarely referred to data, and never to testing a hypothesis, but referred heavily to argumentation. The difference was extreme. These results were significant to $p < 1.9 \times 10^{-53}$. That's: significant to $p < .000000000000000000000000$ $000000000000000000000000000019$, using a chi-square test.

Attribution: Abby Hafer

Permission: This chart was created and is owned by Abby Hafer

CHAPTER 29: THE PUBLISHING SCANDAL THAT ROCKED THE DISCOVERY INSTITUTE

Photocopies: Two pages of the 2007 tax forms from the Discovery Institute

1) Showing Richard Von Sternberg's salary and expenses at the Discovery Institute

2) Showing Stephen Meyer's salary and expenses at the Discovery Institute

These tax returns were obtained through the Freedom of Information Act. They were given to the author by Dr. Barbara Forrest.

Highlighting done by Gretjen Hargesheimer

CHAPTER 30: BAD DESIGN—SHARKS GET MORE REPRODUCTIVE OPTIONS, TOO.

Screenshot: TV news footage of a great white shark savaged by another shark, probably another great white.

Screenshot caption: Some sharks have poor social skills.

Original title: None. This was a newscast.

Source: http://www.youtube.com/watch?v=nnokWWyGk3A

Gray scaling and cropping by Gretjen Hargesheimer

CHAPTER 31: EXPLODING THE CAMBRIAN EXPLOSION

Photograph: An Ediacaran fossil called *Dickinsonia costata*.

Original title/description: DickinsoniaCostata

Attribution: Original uploader was Verisimilus at en.wikipedia

Copyright information: This file is licensed under the Creative Commons Attri-bution-Share Alike 3.0 Unported license.

Gray scaling and cropping by Gretjen Hargesheimer

CHAPTER 34: BAD DESIGN—THE HUMAN APPENDIX

Illustration: Human appendix and colon.

Original drawing by: Alexander Winkler

Copyright: Abby Hafer

Illustration: Adult human female pelvis.
Original drawing by: Alexander Winkler
Copyright: Abby Hafer

Illustration: Human hair with *arrector pili* muscle.
Original drawing by: Alexander Winkler
Copyright: Abby Hafer

Framed quote: Alfred Sherwood Romer and Thomas S. Parsons, *The Vertebrate Body,* 6th ed. (Philadelphia: Saunders College Publishers, 1986), 389.
Illustration by: Alexander Winkler

CHAPTER 35: EVOLUTION: THE GREATEST INDISPUTABLY TRUE STORY EVER TOLD

Photograph: Our place in the galaxy.
Original title/description: This picture taken with ESO's Very Large Telescope shows the galaxy NGC 1187. This impressive spiral lies about sixty million light-years away in the constellation of Eridanus (The River). NGC 1187 has hosted two supernova explosions during the last thirty years, the latest one in 2007.
Attribution: European Southern Observatory Very Large Telescope
Copyright information: ESO images and videos, along with the texts of press releases, announcements, pictures of the week and captions, are released under the Creative Commons Attribution 3.0 Unported license and may on a non-exclusive basis be reproduced without fee provided they are clearly and visibly credited.
Arrow and "You Are Here!" label added by Gretjen Hargesheimer/Abby Hafer
> Note: Since this is not a photograph of our Milky Way galaxy, the viewer needs to understand that the arrow does not indicate literally where Earth is in that galaxy, because Earth is not in the NGC 1187 galaxy. The arrow indicates the approximate position of the earth (about half way out, on a spiral arm) in the Milky Way galaxy, which is shaped similarly to NGC 1187 .
Gray scaling and cropping by Gretjen Hargesheimer

Chart: A phylogeny of everything (phylogenetic tree).

Chart caption: Or, our place on the family tree.

Original title/description: Universal phylogenetic tree in rooted form, showing the three domains. According to C. Woese et al. 1990

Attribution: Maulucioni

Based on Woese et al. rRNA analysis http://commons.wikimedia.org/wiki/File:Current_location_in_evolution.png

Copyright: Public domain

The chart for this book was done by Gretjen Hargesheimer.

Arrow and "You Are Here!" label added by Gretjen Hargesheimer/Abby Hafer

Gray scaling and cropping by Gretjen Hargesheimer

APPENDIX 1: THE PHYLOGENY OF MAMMALIAN TESTICLES

Diagram: Phylogeny of Mammalian testicles.
Diagram caption: The evolution of the scrotum and testicular descent in mammals: a phylogenetic view.

Source: Werdelin, L, Nilsonne, A. Journal of Theoretical Biology 1999 Jan 7;196(1): 61–72. Department of Palaeozoology, Swedish Museum of Natural History, Box 50007, S-104 05 Stockholm, Sweden. werdelin@nrm.se

Original diagram by: Alexander Winkler

Copyright: Abby Hafer

After L. Werdelin L and A. J. Nilsonne, "The evolution of the scrotum and testicular descent in mammals: a phylogenetic view," *Theor Biol.*, January 7, 1999, 196 (1) 61–72.

APPENDIX 3: THE LIFE CYCLE OF THE IMMORTAL JELLYFISH

Illustration: The repeating life cycle of the immortal jellyfish (*Turritopsis nutricula*).

Original drawing by: Alexander Winkler

Copyright: Abby Hafer

After: Transformation pathways of sexually mature (14–16-tentacle) medusae of *Turritopsis nutricula* in Scott F. Gilbert, *A Companion to Developmental Biology,* 8th ed., Sinauer Associates, http://9e.devbio.com/printer.php?ch=&id=6

The original research is reported in: S. Piraino, F. Boero, B. Aeschbach, and V. Schmid, "Reversing the life cycle: Medusae transforming into polyps and cell transdifferentiation in *Turritopsis nutricula* (Cnidaria, Hydrozoa)" *Biol. Bull.* 1990, 90: 302–12.

Index

Index

appendix. *See* human appendix
appetite-suppressing hormone, 155
aqueducts, Roman, 84
Arca barbata (bivalve), 107
Archeopterix (bird-like dinosaur), 78
arrector pili muscle, 180
art imitating nature, 85
astronomy, 37

B

baby teeth, human, 144
Bachmann, Michelle, 142
bacteria
  and antibiotic resistance, 155, 183
  and evolution, 63
  and flagella, 100
  Helicobacter pylori and ulcers, 154
  in the human appendix, 178
  in our mouths, 145
  photosynthetic, 165
bacterial diseases treatable by antibiotics, 58
"bacterial flagellum," 9, 100–101
bad design in humans
  appendix, 177–81
  biochemical pathways (vitamin C),
    126–27, 132–35
  birth canal, 45–51
  blood clotting system, 91–96, 97–99
  eyes, 111–15
  the genome (and mutations), 171–76
  reproductive options, 162–63
  teeth, 144–47
  testicles, 13–15
  throat, 71–73
the "bad old days," 130
balls. *See* testicles
Bangladesh, mortality in childbirth, 51
beauty of the human body, 182
behavior and its influence on the afterlife, 122–23
Behe, Dr Michael
  credentials, 22
  age of the earth, 21
  common descent, 21
  irreducible complexity, 9, 81, 101–2
  on malaria, 119

*Edge of Evolution: The Search for the
  Limits of Darwinism* (book), 101
the benefits of science, 43, 57
beta-thalassemia, 175
better body parts. *See* animals in which
  some parts are better than ours
the Bible
  biblical advice, 124
  biblical creationism, 19
bigger killers than AIDS, 117–19
biochemical pathways (vitamin C)
  definition, 127
  discussion, 126–35
  humans vs cats, 130
  in rats, 134
  in seals, 132
  in sled dogs, 132
biological family tree, roots, 166
the Biological Society of Washington,
  157, 160
biology, evolutionary and evidence for
  evolution, 155
biology textbooks, stickers, 54
bird testicles. *See* testicles
birth canal, 45–52
bivalves
  Arca barbata, 107
  Cardium muticum, 105
  Lima squamosa, 106
  Pecten maximus, 105
  Tridacna maxima, 106
blasphemy. *See* crimes against dogma
blind spot, 111–13
  See also eyes
blood clotting disorder (hemophilia),
  97–99
blood clotting system, 91–96, 97–99
body temperature and sperm production, 3–4, 13–15
  See also testicles
British sailors and scurvy, 130
bubonic plague, 58
bumblebees, ability to fly, 168

C

Caesarian section, in childbirth, 46, 49
Calvinism, 123–25

Index

Cambrian geological period
  the Cambrian Explosion, 164–66
  and evolution, 166
  pre-Cambrian fossils, 20, 164–65
  Cambrian fossils, 20
  post-Cambrian organisms, 166
cancer
  carcinogens, 175
  and mutations, 173, 174–75
  new treatments, 155–56
  and tobacco, 60, 136–37, 140
Carcharias taurus (sand tiger shark,
    gray nurse shark), 163
Cardium muticum (bivalve), 105
Carson, Ben, 142
Catholicism, 122
  See also Christianity
cats, and vitamin C, 130
cattle feed. *See* animal feed and
    antibiotics
cavities, dental, 145
Center for Science and Culture
  and HIV/AIDS denial, 117
  and Intelligent Design, 8
  personnel, 18, 20–22
cephalopod eye, 115
cervical cancer, 57
changing the subject, as tactic, 139
childbirth in humans. *See* birth canal
cholera, 58, 119
cholinesterase deficiency, 175
choroideremia, 175
Christianity
  and morality, 122–23
  conservative Christian viewpoint, 7,
    10, 26
  loopholes for bad behavior, 122
  religious press, 27
  See also Catholicism
cigarettes and cigars. *See* tobacco use
cilia. *See* flagella and cilia
citrus fruits, as remedy for scurvy, 130
clams (bivalves), 107
Clean Air Act, Clean Water Act, 31
Cobb County, Georgia, biology text-
    book sticker, 54
coccyx (human tail). *See* vestigial
    features

cold-blooded animals, reproductive
    organs, 4
color change in moths, 64
common ancestors
  between humans and chimpanzees,
    21
  and vitamin C, 133
  See also evolution by natural
    selection
Communism
  and the Russian Revolution, 98–99
  in the Soviet Union, 33–35
complexity in biological organisms,
    171
  See also irreducible complexity
complications of childbirth, 48–52
compromise, as a solution, 54
concept of heaven and hell, 122–23
conscious design vs natural phenom-
    ena, 81–90
the consequences of denying science
  agriculture in the Soviet Union,
    33–35
  HIV/AIDS, 117
  polio, 119–20
  Russian history, 98
conservative Christians. *See*
    Christianity
Constitution. *See* U.S. Constitution
"continuing controversy," and evolu-
    tion, 11
  See also manufacturing doubt and
    uncertainty
continuous replacement of teeth, 148
controlled experiment, definition,
    37–38
controversy, manufactured, 53–54
copy errors, in genetic code, 172–75
court cases
  Dover, Pennsylvania school trial
    (2005), 6
  Edwards v. Aguillard (1987), 23
  *Kitzmiller v. Dover Area School Dis-
    trict* (2005), 5, 25
  Scopes trial (1925), 6
the court of public opinion, 11–12
creation ex nihilo, 27

Index

creationism
definition, 20
and Intelligent Design, 17–28
in public schools, 11–12, 23
in textbooks, 23–24
creation myths, 42, 76, 121–22
Cretaceous geological period, 78
Cretaceous-Tertiary extinction, 77
crimes against dogma, 123
crippled children, from polio, 119
"critical thinking," 76
crop yields, in the Soviet Union, 34
crowning, in childbirth, 45
the Crusades, 122
C-section. *See* Caesarian section
cuttlefish, eye anatomy, 114–15

D

Darwin's theory. *See* evolution by natural selection
deaths
AIDS, in South Africa, 116, 117
antibiotic-resistant bacteria, 183
appendicitis, 178
asphyxiation, 72
childbirth, 46, 48, 50–51
dental disease, 145, 147
diptheria, 120
measles, 120
MRSA, 117–18
polio, 119–20
prior to agriculture, 147
scurvy, 130–32
scurvy vs drowning and piracy, 130
the defense of evolution, 184
Dembski, Dr William A
credentials, 18
and the Cambrian explosion, 164
and information theory, 167–70
on Intelligent Design, 26–27
and irreducible complexity, 81
on the Wedge strategy, 32
The Design Revolution (book), 48,
167, 171
Of Pandas and People (book), 26
dentistry and dental problems, 144–47
See also teeth

Dent's disease, 175
denying science. *See* consequences of
denying science
the Designer, 10, 27–28, 77
design inference
definition, 82
and geology, 81–90
See also irreducible complexity
Design of Life (textbook), 26
design proponents vs creationists, 23
The Design Revolution (book), 48, 167,
171
Dickinsonia costata, Ediacaran fossil,
164–65
digesting wood, 178
digestive system. *See* the appendix;
human throat
dinosaurs, 77, 165
diptheria, 57, 120
The Discovery Institute
background, 7–8
stated goal, 31
Intelligent Design funding, 142
publishing scandal, 157–61
scientific research, 151–55
tax returns, 158–59
and the Wedge strategy, 29–31
See also Center for Science and
Culture
diseases
disease-causing organisms, 78
genetic (caused by mutations), 173,
174–75
prevented by vaccines, 57
diversity of opinions, among Intelligent
Design supporters. *See* range of
opinion
DNA
as "signature" of the Creator, 133
See also human genome
dogma, crimes against, 123
dolphins, 14, 85
Dover, Pennsylvania
*Kitzmiller v. Dover Area School Dis-
trict* (2005), 25
politics and the school board, 11
Duesberg, Peter and HIV/AIDS denial,
117

Index

God
    and Intelligent Design proponents,
        7, 27–28
    and William Dembski, 168
    "God of the gaps" strategy, 59
    "God works in mysterious ways," 74, 79
gonorrhea, 58, 119
the "good enough" principle, 16
the "good old days," 125
"goose bumps," 178–79, 180
gravity, and helium balloons, 58
grey nurse sharks (sand tiger sharks),
    163
gum infections, 145–47, 146
    See also teeth

H

haemophilus influenzae (the "flu"), 57
Haiti, and diptheria, 120
"half an eye." *See* eyes
The Handy Dandy Evolution Refuter
    (handbook), 53, 74
heaven, as reward for good behavior,
    122–23
Heimlich maneuver, 72
Helicobacter pylori and ulcers, 154
helium balloons and gravity, 58
hell, as punishment for bad behavior,
    122–23
hemophilia, 97–99, 175
    See also blood clotting system
hepatitis, 57, 119
heresy. *See* crimes against dogma
HIV/AIDS, 31, 63, 78
HIV/AIDS denial, 18, 20, 26, 117
hole in our vision. *See* blind spot
hoodoos (rock formation), 88–89
    See also natural phenomena
horseshoe crabs, 93
    See also blood clotting system
hospital-acquired infections, 118
Hsueh, Aaron, endocrinologist, 155
Huckabee, Mike, 142
human appendix, 177–81
human blood clotting system. *See*
    blood clotting system

human body temperature, and repro-
    duction. *See* body temperature
    and sperm production
human eye, 103–5, 111–15
    See also eyes
human genome (and mutations)
    as bad design, 171–76
    genetic diseases, 173, 174–76
    and Dr Dembski, 168
Human Immunodeficiency Virus. *See*
    HIV/AIDS
human papillomavirus, 57
humans and chimpanzees, common
    ancestor, 21
human tail. *See* vestigial features
human throat, 71–73
Huntington's chorea, 173
hypothesis and prediction
    definition, 37
    testing, or lack thereof, 153
hyraxes, 14

I

ID. *See* Intelligent Design
ideology over evidence, in the Soviet
    Union, 34–35
IFT (interflagellar transport), 101
image processing, in humans, 114
    See also eyes
immortal jellyfish, 69–70
immune-suppressed patient, and
    measles, 120
improbable events, 160–61
inactivated poliovirus vaccine, 119
    See also polio
infection in jaw, teeth or gums, 145–47
    See also teeth
infections, previously curable, 118
influenza, 57
information theory
    definition, 167
    and Dr Dembski, 167–70
    information theorists, their opinion
        of Dr Dembski, 169
innoculation against smallpox, historic
    opposition to, 56
Intelligent Design (ID)

Index

Lysenko, Trofim
  theory of acquired characteristics,
    33–34
  and agriculture in the Soviet Union,
    35

M

"macroevolution," definition, 60
Mafia members, excommunicated, 122
malaria
  malaria parasite, 101
  drug resistence, 119
  Dr Behe on, 101–2
male testicles. See testicles
mammals with convertible testicles, 13
manatees, 14
manufacturing doubt and uncertainty,
    53–56, 136–37, 140
Mark Twain, as example, 75
mass extinctions, 77–79
"materialism"
  definition, 7, 8, 30
  and the Wedge document, 30, 125
Mbeki, Thabo, President of South
    Africa, and HIV/AIDS denial,
    116–17
measles, 57, 120
measurement and testing, 37, 64
the media and Intelligent Design, 142
medical breakthroughs, 155–56
megalodon (C.megalodon), and teeth,
    149
meningitis, 58, 119
men's reproductive organs. See testicles
mental illness, as possesion by demons,
    56
Methycillin Resistant Staphylococcus
    aureus. See MRSA
Meyer, Dr Stephen
  on DNA, 133
  compensation from Discovery Insti-
    tute, 157–59
  scientific paper, and its retraction,
    157, 160
"microevolution" vs "macroevolution,"
    59–61, 63–64
miscarriage, in pregnancy, 51

Missouri
  legislation regarding teaching Intel-
    ligent Design, 11–12
  polio outbreak (1979), 119
mitochondria, 172
modern medicine vs religious rites, 121
Moon, Reverend Sun Myung, 19
moral codes and religion, 42–44,
    121–25
Morris, Simon Conway, paleobiologist,
    20
mortality. See deaths
moths, color change, 64
MRSA (Methycillin Resistant Staphylo-
    coccus aureus), 117–18
mudskippers, 65–68
multicellular organisms, fossils, 164
mumps, 57
mutations
  definition, 173
  and cancer, 174–75
  consequences of, 175
  and evolution, 174
  and vitamin C, 127–34
  See also human genome

N

natural phenomena
  natural arches, 82–85
  vs conscious design, 81–90
  See also irreducible complexity
natural selection. See evolution by
    natural selection
Nelson, Dr Paul A
  credentials, 20–21
  and the Center for Science and
    Culture, 26
  and Young Earth creationism, 25
the Netherlands, polio outbreak
    (1992–93), 119
neurofibromatosis (genetic disease),
    175
new species
  definition, 60, 63
  and Intelligent Design, 6, 78
  new microorganisms, 63–64
  new disease-causing organisms, 78

Index

Index

Index

the uniqueness of human beings, 183
United States, settlement, 124
unvaccinated population, consequences, 119–20
U.S. Constitution, Establishment Clause, 23
U.S. Supreme Court, rulings on creationism and Intelligent Design in public schools, 23, 25
"utopianism" and science, 31, 125
See also "materialism"

V

vaccines
    and germ theory, 40, 56, 57
    vaccine-preventable diseases, 57–58
    vaccination campaigns, 120
    vaccination refusal, 119–20
varicella (chickenpox), 57
variola (smallpox), 57
Vendian fauna. *See* Ediacaran fauna
vermiform appendix. *See* human appendix
vertebrates, blood clotting system, 95
vestigial features
    human appendix, 178
    human tail, 179
    human arrector pili muscle, 180
    See also appendix
victimless sins, 123
viral hepatitis, 119
Virginia, legislation regarding teaching Intelligent Design, 11
vision. *See* eyes
vitamin C. *See* biochemical pathways (vitamin C)
Von Sternberg, Richard, 157–58
    See also publishing scandal
voyages of exploration
    death rate, 130
    and scurvy, 134

W

Walker-Warburg syndrome (genetic disease), 175
walking fish. *See* mudskippers

walking upright, and childbirth, 46
warm-blooded animals and sperm production, 13–15
    See also testicles
wars and religion, 122
the Wedge strategy
    goals of Intelligent Design movement, 8, 29–32
    leaked in 1999, 29
    as a liability, 32
    and "utopianism," 125
    The Wedge Strategy, 1998 (document), 29–30
weight of scientific evidence, 55
Wells, Dr Jonathan (John Corrigan Wells)
    credentials, 20
    background, 19
    and HIV/AIDS denial, 20, 26, 117
whales, 14, 73
"What good is half an eye?" as tactic, 103, 106
wheat in Soviet Union, 33–35
who is promoting Intelligent Design?
    Ahmanson, Howard F. Jr, 7
    Behe, Dr Michael, 22
    Dembski, Dr William A, 18
    Johnson, Phillip E., 17–18
    Nelson, Dr Paul A, 20–21
    Wells, Dr Jonathan (John Corrigan Wells), 20
    See also under each name
width of women's hips, and childbirth, 47
wild polio viruses. *See* polio
Wisconsin, polio outbreak (1979), 119
women's pelvis. *See* birth canal
wood, digesting, 178
writing styles, rhetorical vs scientific, 152–53

Y

yellow fever, 57
You Are Here!, 184, 185
Young Earth creationism, 20, 25, 77
young people, as message targets, 139, 143

#0011 - 231018 - C0 - 229/152/13 - PB - 9780718894207